Why Leather?

Sidestone Press

Why Leather?

The material and cultural dimensions of leather

edited by
Susanna Harris
& André J. Veldmeijer

Published by Sidestone Press, Leiden
www.sidestone.com

ISBN 978-90-8890-261-1

Lay-out & cover design: Sidestone Press
Photograph cover: Vitaly Korovin | Dreamstime.com

Contents

Preface

Susanna Harris & André J. Veldmeijer

The chapters in this volume originate from the conference 'Why leather?' organised by the Archaeological Leather Group (ALG) in 2011. The choice of theme arose from a discussion between members of the Group who recognised that, while it is possible to generalise about certain aspects of leather, the variability in materials that come under this category and corresponding attitudes to it, demonstrate a wide variety of materials and associated values.

The call for papers acknowledged that while few doubted the common occurrence of leather through time, the question of why people use leather, or more specifically certain types of leather, raises a number of issues. Participants were asked to question the material properties of animal skins and their products, to question their qualities, and their desirability or otherwise. They were asked to consider the origin of leather, the values attached to it and the interchangeability of leather with other materials such as textile, paper and plastic. Participants were asked to question the relationship between crafts workers and supplies, technique and design, practicality, ethics and value. The conference provided the opportunity to explore these material and cultural dimensions of leather by encouraging participants to address the question 'Why leather?'

The authors in this volume come from a rich pool of talent, from crafts workers, professionals in various spheres of the leather industry, materials engineers, to archaeologists and historians specialising in the study of leather. A short profile of each is provided in the section 'About the Authors'. Separately, the contributors bring their individual knowledge, skills, experience and interests to the topic. Together, they provide an innovative and thoughtful reply to the conference question.

The conference was organised and funded by the ALG. The Group's income mainly derives from its member's annual subscription fee and we hope they are happy with the way their money has been spent. We are grateful to the World Archaeology Section, Institute of Archaeology, University College London for providing the venue for the conference. Thanks to the British Academy for their financial support of Susanna Harris' Post-doctoral Fellowship 'Cloth Cultures in Prehistoric Europe', during which period this conference was organised. We acknowledge with gratitude our publishers, Sidestone Press, for their enthusiastic response to this publication and their fabulous publishing model which allows free online-access, low-cost PDFs as well as publish on demand paper copies, of all the chapters herein. Thanks to those individuals who provided advice and support from initial conception to publication: Caroline Cartwright, Diana Friendship-Taylor, Margarita Gleba, Jackie Keily, Katie Meheux, Quita Mould, Lucy Skinner, Roy Thomson, Barbara Wills and Sue Winterbottom. Special thanks to Roy and Pat Thomson for copy-editing the volume and Adri 't Hooft for type-setting.

London/Dwingeloo/Cairo
26 June 2014

About the Authors

Eddie Cheshire, MSc, PhD served a four year mechanical and production engineering apprenticeship working on aircraft engine accessories. He has since spent over 35 years in industry, working mainly with composite materials such as glass fibres and polyester resin. The typical products Cheshire worked with were large diameter sewage pipes, the machinery to make telephone poles, corrugated roof sheet and the side panels of refrigerated lorries. In 1990 he studied for an MSc in Composite Materials at Imperial College. For nearly 25 years he has been a Residential School tutor for the Open University. Cheshire retired from industry in 2002 and carried out research on *cuir bouilli* and other non-metallic armour, leading to a PhD from the University of Reading in 2010. His industrial background is particularly relevant because *cuir bouilli* is a composite material reinforced by fibres of collagen.

Susanna Harris, BA (Hons), MA, PhD is Research Associate at the Institute of Archaeology, University College London. She is interested in the materials and technology of cloth and clothing from the Mesolithic to Iron Age in Europe. Her research focuses on textiles, leather and basketry, and the relationship between these cloth-type materials as cloth cultures in past societies. Harris' approach combines analytical, theoretical and experimental approaches to artefact studies. Recent publications include 'Sensible dress: experiments with the sight, sound, touch and smell of Late Ertebølle, Mesolithic cloth types' (Cambridge Archaeological Journal, 2014) and 'From the parochial to the universal; comparing cloth cultures in the Bronze Age' (European Journal of Archaeology, 2012).

Dr. Salima Ikram is Professor of Egyptology at the American University in Cairo, and has worked as an archaeologist in Egypt since 1986, directing the Animal Mummy Project, and co-directing the Predynastic Gallery project, the Egyptian Museum Chariot Project, and the North Kharga Oasis Survey. Ikram has lectured and published extensively, both for children and adults.

Amanda Michel BSc (Hons), FSLTC, has been involved with many innovative, yet practical, scientific research projects in leather for more than 35 years, enabling her to become specialised in the microscopic structure of leather and how it influences its behaviour in use. She has been involved with virtually every aspect of the leather business – from the animal on the farm, through each stage of the leather making process, to product design and manufacture. She also has extensive understanding of the consumers' perception of the final product. This experience enables her to deliver training and consultancy in all aspects of leather and its uses to anyone in the supply chain. In addition to having published many papers, Michel regularly writes technical articles for the leather trade press and was the author of a chapter on leather microscopy in the Society of Leather Technologists and Chemists' publication, The Leather Technologists Pocket Book. She is a Fellow of the Society of Leather Technologists and Chemists and is currently President of the Society. Since October 2005, Michel is the director of her own leather consultancy and training business, Leather Wise Ltd.

Alan S. Raistrick was brought up near Bradford in a family of tanners making leather for use on the machinery in the textile industry. He remembers looking round the family tannery at about five years old, and a textile mill at about seven. After studying chemistry at Imperial College he did research for the leather industry for 15 years before becoming a computer engineer. He was in charge of the Imperial War Museum film vaults for a couple of years. After retiring early he researched both the history of leather making and the textile machines he had made leathers for during school and university holidays. His late wife was a skilled hand spinner which, coupled with his historical interests, led to a particular interest in preindustrial spinning wheels and hand looms.

Amanda Watts, BA, MA, MSc studied archaeology and fine wine at Boston University, USA, before pursuing a career in archaeological conservation, earning her MA and MSc in object conservation at the Institute of Archaeology, University College London. She currently works as an archaeological conservator at the Mes Aynak excavations in Afghanistan. Recent projects include an assessment of the condition and exhibition considerations for human remains at the Petrie Museum of Egyptian Archaeology, London. Publications include a review of the use of glycerol treatment on archaeological leather with the Museum of London Archaeological Conservation Department and the London Archaeological Archive and Research Centre (LAARC). Over the years between archaeology and conservation, Watts has had a parallel career as a wine steward, vineyard representative and professional wine taster.

Barbara Wills works as a conservator at the British Museum London on a range of objects made from organic materials, specialising in the conservation of leather, basketry materials, human remains and ancient Egyptian objects. She recently completed a Senior Clothworkers' Conservation Fellowship, undertaking a project to study and stabilize a group of natural mummies from recent salvage excavations near the Fourth Nile Cataract in Sudan. She is a Trustee of the Leather Conservation Centre and an accredited member of the Institute of Conservation.

Dr. André J. Veldmeijer is Assistant Director for Egyptology at the Netherlands-Flemish Institute in Cairo. He has worked as an archaeologist in Egypt since 1995, specialising in leather and footwear among other topics. He (co-)directs several projects such as the Ancient Egyptian Leatherwork Project, which incorporates the Egyptian Museum Chariot Project, and the Tutankhamun Sticks and Staves Project. He has published extensively including both scientific and popular literature.

Laura Youngson Coll BA (Hons), MA initially trained as a sculptor and installation artist gaining her Bachelor's degree in 2001 and Masters in Sculpture at the Royal College of Art in 2004. Youngson Coll subsequently worked with bookbinders and leatherworkers for many years, learning traditional techniques, gaining the 10,000 hours synonymous with a skilled craft practice and making bespoke luxury interiors. In 2012, she set up her own practice taking these skills and exploring their potential within her own sculptural work. In 2012, Youngson Coll was shortlisted for the Cockpit Arts/Leathersellers Award and was recently commissioned by Craftspace, Somerset Art Works and the National Trust to make site specific work at Barrington Court for the 'Make the Most' exhibition. In 2014, she was winner of the prestigious Perrier-Jouët Arts Salon Prize.

Introduction

Leather in archaeology: between material properties, materiality and technological choices

Susanna Harris

Why leather?

The study of leather is a specialist field in archaeology, yet focuses on one of the major materials in the past, the use of which continues into the present. The common occurrence of these animal skin products through time, whether tanned leather, parchment, vellum, oil or fat cured skins or rawhide attest to the enduring utility and desirability of animal skins as a material. Traditionally, these products have all been grouped together as leather (Hodges 1995, 151), although their fundamental differences are increasingly recognised and published in the archaeological literature (Thomson 2006a, 1-3). For those without a specialist interest in leather it is easy to overlook the variability of products among this group of materials and to lose sight of the specific reasons behind the choice of leather in particular situations and according to different cultural and temporal contexts. In this volume, the authors address the question 'Why leather?' through investigating the nature of animal skins, the behaviour of skins and leather in use and the network of decisions made by the makers, designers and users in bringing raw materials to a finished object and its place in the social fabric of life. The authors also address why leather works in certain situations, and indeed sometimes why other materials were and are chosen in preference to leather. The response to such questions is not only addressed through the properties of materials, but also how leather, like all materials, is viewed with the dimensions of culture and beliefs which surround it. The aim of this introduction is to place the conference theme and chapters herein within the field of leather in archaeology and current issues surrounding the study of materials in the past. This volume benefits from the insights of archaeologists and authors from other professions, whose specialist knowledge provides the archaeologists with a new perspective of seeing their topic.

Skins to leather

An animal skin, left without treatment for several days, will quickly decay. A simple intervention of removing the subcutaneous fatty layer and drying will reduce the action of harmful bacterial and slow down the rate of decay. It is only through a more lengthy process that animal skins are cured, tanned or dressed to create a stable product (see Hodges 1995, 148-52; Reed 1972, 46-86; Sharphouse 1983). These processes lead to different products. In English the word 'leather' is used to refer to a wide range of animal skins products, indeed it is used in this manner in the question 'Why leather?'. Technically, however, leather should only refer to animal skins which have been "rendered non-putrescible under warm moist conditions" (Thomson 2006a, 3). Parchment and vellum[1], for example, are stretched and dried skins which are unstable under warm, moist conditions. Buckskin refers to fat cured skins of variable stability, while true leather processed by vegetable, chrome or other tanning techniques is what we are most familiar with in our shoes and handbags (for details see Thomson 2006a, 2). Such classification and English language terminology reflects differences in raw material, process and product; no doubt all leather producing peoples had their own language to distinguish different animal skin products.

The enduring presence of leather may be attributed to its properties as a strong, flexible, sheet material and its ready availability in cultures where animals are slaughtered for meat. Yet, the finished product depends greatly on the nature of the raw material and the way it is processed. Leather can be soft and supple like a textile, firm and rigid like a basket, or hard and watertight like a pot or gourd. The varied properties of leather are in part due to the chemical and physical composition of animal skins (Haines 2006a), the species or breed from which it originates (Haines 2006b, 12-19), and the method with which it was treated (Covington 2006; Thomson 2006a). For the tanner, an understanding of the way animal skins transform during processing allows them to create a particular material. For the craftsperson, an understanding of leather as a material allows them to create and innovate according to the desires of the designer and consumer. To the consumer, a particular material may be desirable due to its practicality, aesthetic, novelty or availability.

Yet, a utilitarian or functional approach to material properties requires careful questioning; in a world of technological and material choices, archaeological theorists recognise that people appropriate materials according to their suitability on many levels (Sillar & Tite 2000, 3-9). The separation of materials into functional or symbolic is criticised as a simplistic and unhelpful dichotomy (Meskell 2005, 2). In recent years, it has been proposed that mind, agency and matter are co-dependent and come together in the object (Knappett 2005, 85). Accordingly, we should no more prioritise the empirical analysis of material properties, than the meaning of leather as the essence of creativity and human agency, but see these as integrated aspects surrounding the very actions and choices of making and using materials in daily life.

---

1 Parchment is made from sheep or goat skin, whereas vellum is made from calf.

In this volume the chapters follow a range of perspectives within this sphere; however, the extremes of either approach are tempered by the recognition of the relationship between the material and cultural dimensions of leather. The authors show why some leathers work and are perceived to work in certain situations, and why these materials are chosen over other materials, or other types of leather. In several chapters the approach is very much centred on the internal dynamic of the material. Amanda Michel demonstrates how the structure of skins, as viewed microscopically, influences the behaviour of skins and certain species of skins in use, while Alan Raistrick provides an account of the leathers with specific performance characteristics (thickness, smoothness, grip, strength) for use in nineteenth and twentieth-century textile machinery. Here, it was not only species (cow, goat) that mattered, but also the age, breed, climate and way the animals were cared for.

Yet even among such materials based selection, the authors recognise a more holistic approach to leather. Michel has to understand the consumer's perception of the final product, while Raistrick notes what he calls a 'Samson Complex', as the customer believed (wrongly, in Raistrick's opinion) that hair-on leather bands were stronger. Similarly, there were requests for coloured leathers in textile machinery, which again in Raistrick's opinion were more to do with a memory of the efficacy of former coloured leather, than actual performance. By contrast, leather craft practitioner Laura Youngson Coll points out the discrepancy between design brief and materials, where specialist leathers are proposed and used for purposes that are not necessarily suitable for those materials: leather floors in stores where women wear stiletto heels or leather covered plant pots. Here, factors other than utilitarian properties are leading the selection of material. These points show the complex relationship between mind, agency and matter which is found in the selection of materials and not readily explained by a strict empirical approach.

Leather in archaeology

In archaeology, the answer to the question 'Why leather?' is complicated by the poor survival of leather and other skin based products and the difficulty of identifying their origin and the way they were processed. When leather is preserved, its contact with the preservation environment often means it has undergone chemical and physical transformations. These changes may hinder the identification of the animal species from which skins originate, and make it difficult to determine the method by which the animal skin was processed. Species identification by microscopic techniques depends on the preservation of morphological features of the skin or hair, which may be seriously degraded (Appleyard 1978; Leather Conservation Centre 1981; Teerink 1991; Wildman 1954). When preservation of sufficient molecular structure allows, DNA or protein mass spectrometry techniques (also referred to as ZooMS, collagen or peptide fingerprint, protein sequencing) offer new means of species identification (Collins *et al.* 2010; Hollemeyer *et al* 2012; Schlumbaum *et al.* 2010). In terms of archaeological evidence, leather processing may be detected through the preservation of associated tools (for example, Thomson & Mould 2011; Mould *et al.* 2003; Ottaway & Morris 2003; Raedler

2007; Schwarz 2002), evidence of the substances applied to skins (Rifkin 2011), and chemical or biological analysis of organic residues remaining within the leather itself (Driel-Murray 2002; Falcão & Araújo 2011; Groenman-Van Waateringe *et al.* 1999, 886-890; Thomson 2006b, 58-59). Archaeologists must draw on as many lines of evidence as possible to build up an understanding of leather in the past.

Poor preservation contrasts with the notion that leather is thought to have been a common material in the past. Since humans migrated to the northern hemisphere, it has been assumed they wore skins to protect themselves from the cold, an assumption supported by the ubiquity of stone scrapers, cut marks on animal bones associated with skinning in the Palaeolithic (Charles 1997) and the early presence of lice whose habitat is clothing worn by humans (Toups *et al.* 2011). In Europe and the Mediterranean it is only from the Roman period onwards that leather becomes more common in archaeological excavations, most likely due to the method of processing leather by vegetable tanning (Driel-Murray 2000, 305; Groenman-Van Waateringe *et al.* 1999, 885-886, 889-890). In some Medieval and Post-Medieval urban centres with waterlogged deposits, leather finds are sufficiently common that all but the most exceptional leather small finds are classified as a bulk (Grey 2006, 28). In this volume André Veldmeijer and Salima Ikram present the remarkable remains of the only known complete example of the leather casing, harnessing and related leather equipment of a New Kingdom Egyptian chariot, preserved for over 3000 years in the dry environment of Egypt. Through comparison with surviving wooden chariots, they are able to draw conclusions about physical and aesthetic properties of the chariot materials, for chariots that were used either in lavish processions or in the fast paced heat of battle. Where leather is not preserved, archaeologists work with other sources of evidence. Using written, pictorial and artefactual sources, Barbara Wills and Amanda Watts investigate early evidence for wineskins in the Mediterranean and consider this in relation to historical knowledge of wine transported in wineskins via mule trains. From this evidence they are able to build up a vivid picture of the qualities of the wineskins, the taste the skins imparted to the wine, the riotous feasting and drinking, along with godly behaviour that was related to wineskins. In these chapters the authors addresses the question 'Why leather?' from the perspective of materials integrated into daily life with its varied complexity.

Assessing materials

There are several routes available to the archaeologist to understand how materials, such as leather, may have performed in use. One route is the chemical and physical testing of raw materials (animal skins) or its processed form (leather and other skin products); these methods are based in the natural sciences. The other route is the comparison of contemporary or historically known leather types and leather objects with those under investigation, on the assumption that past materials performed in similar ways. The comparative approach uses sources from experiment and design, ethnography and history. The natural sciences and comparative approaches may be used singularly or in combination, and these different methods can be directed towards understanding diverse aspects of leather.

As introduced above, Michel uses her extensive knowledge of the microscopy of animal skins gained through working in the leather industry to evaluate how the structure of animal skins influences the behaviour of leather in use. She demonstrates the complex interplay between the role of collagen fibre bundles (animal proteins), the structure of the skin according to the area of the hide, and the influence of species. Taking the butt area of the animal, for example, we learn that its compact structure means it is one of the strongest areas of a hide, in comparison to the belly which is stretchy, or the neck area, which is thick but wrinkly. If we then take into consideration species, we are able to understand that wool sheepskins are prone to splitting due to the presence of a layer of fat cells, unlike goat and cattle skins.

Eddie Cheshire uses quantitative tests to assess the efficacy of different recipes for *cuir bouilli* against arrow penetration. According to medieval literature, this form of leather was used to make armour. By their nature, such tests isolate specific features – in this case the extent to which an arrow can penetrate through the armour into the wearer's flesh beneath. As ancient leather is rarely preserved and, when it is, the leather is degraded, fragile and no-longer retains its original qualities, the tests are carried out on modern leather. In this case, the leather was made according to recipes and techniques Cheshire deemed likely for the period. The tests presented are useful in that they provide quantitative results, and hence allow the comparison of specific performance characteristics across different materials or versions of similar materials. Cheshire tests a range of possible recipes by which *cuir bouilli* may have been prepared for use in armour. His results suggest that the *cuir bouilli* was most likely not boiled leather, as previously believed, but boiled rawhide, which was potentially improved by the addition of an applied, hard surface.

Leather is not only renowned for its physical and chemical properties, it also has aesthetic appeal. Recently archaeologists have been increasingly interested in the aesthetic or visual qualities of materials, situated more widely in a full sensory engagement. This sensory approach to materials recognises that people engage with materials primarily through their bodily senses such as touch, sight, sound, taste and smell (for example Delong *et al.* 2012; Edwards *et al.* 2006). A sensory approach to materials has a small but growing appreciation in archaeology (Hurcombe 2007; MacGregor 1999), including experiments on leathers and furs (Harris 2014). In this volume, Youngson Coll shows the attention that craft workers pay to the sensory nature of leather. For example, as a wall-covering material vellum is appreciated for its smoothness to the touch, while visually it is appreciated for its translucency and texture created by the pattern of hair follicles. The skin of the stingray, referred to as 'shagreen', is appreciated for the opulent, glossy surface created by the polished, calcified papillae within the structure of the leather, which can be dyed any colour according to the demands of fashion.

Historical, ethnographic and experimental accounts of leather craft and industry provide a rich resource for archaeologists to investigate the varied ways skins are or were processed, and how the qualities of such materials can be understood through their application and use (some of many sources: Angus 2002; Douglas 1956; Kellogg 1984; Klokkernes 2007; Mason 1891; Oakes & Riewe 1996; Paine 1994, 19-20, 30-41; Rahme & Hartman 2001; Richards 2004; Wilder 1976). Such accounts also provide examples of the way materials and objects are integrated into society, through the organisation of production, the relationship between humans and animals, variations in process and the way the resulting materials are used in specific circumstances. There is much to learn, for example, from accounts of the north European skin working traditions, where women selected the area of hide in combination with the processing method to make shoes that protect feet from either the wet of autumn or the cold of winter (Brandon-Cox 1969, 124) and which were also a source of pride to their wearers. In this volume, Wills and Watts use historical observations of the transportation of wine in pitch-lined goatskins from nineteenth-century Cyprus to understand how the flavour of wine was tainted through this means of transport. Although an exact comparison cannot be made, it does provide information that would be unlikely to be considered without reference to such sources.

Materials and materiality

Materials have long been the focus of archaeologists' theories of the past. However, whereas Thomsen's Three Age System (Stone Age, Bronze Age, Iron Age) was developed to classify artefact collections, and empirical approaches attempt to attach fixed lists of properties to materials, current theories seek to understand the complex interplay between artefacts and the human agent. A recent approach has been to consider objects and materials from the perspective of materiality. Dictionary definitions of materiality focus on the physical qualities or characteristics of the material. This is the way it tends to be used by artists and craftspeople, as for example Youngson Coll who refers to the materiality of leather in terms of her understanding of the way it can be worked. However, the discussion of materiality in archaeology focuses on the relationship between people and things (DeMarrais *et al.* 2004, 2; Meskell 2005, 2, 4). Materiality, as such, emphasises the importance of the relationship between people, objects and materials.

Materiality has proved rather a nebulous concept. To some, it is explained as "how the very material character of the world around us is appropriated by humanity", whereby material culture shapes human lives and relationships (Graves-Brown 2000, 1). To others, this process is expressed as a mutual relationship; materials are shaped by humans, humans are shaped by materials and hence are part of a material world which cannot be separated from social practices, as both bring each other into existence (Jones 2004, 330). From another perspective materiality encompasses the ideas embedded in an object in specific contexts. Here materiality is to materials what gender is to sex; in other words a social construct grounded in a body or object (Hurcombe 2007, 537). Although there are many approaches to materiality in archaeology, the broad underlying theme is the desire to address the interrelationship between the physical materials (their qualities and properties), the perception, use and appropriation of materials, and how these are part of historically situated human lives. These approaches specifically seek to reject what is seen as a false dichotomy between objects as either highly symbolic or purely functional (Meskell 2005, 2). For some scholars, materiality stands in contrast to a materialist approach which places more emphasis on the significance of material properties and material attributes. However, this is distinct from a strictly empirical approach, as materialist approaches recognise that material properties are not fixed so much as relational and processual, as materials are embedded within histories and the 'currents of the lifeworld' and therefore vary according to situation (Ingold 2007, 1-3, 9) (for position papers see Ingold 2007; 2012; Jones 2004; Tilley 2007).

This makes generalisations about material problematic. On the one hand, archaeologists seek to identify materials, such as leather, and processes, such as tanning, in a highly empirical manner to allow them to draw generalisations about material properties and use. On the other hand, the common occurrence of leather or a certain type of leather, does not mean that it fulfilled the same role, was associated with the same ideas, or shared all perceived qualities in all times and places, by all people. Indeed, the rejection of leather and furs by certain sectors of the population over the last few decades attests to the complex interplay of motivations and emotions surrounding materials (for example, Harper 2008). Youngson Coll addresses leather from several of these perspectives. In the world of interior design there are discernible fashions in materials; the showy opulence of vellum and shagreen is highly desirable to some clients at the time of writing. Such issues are complicated by concern for animal welfare and environmental issues: the sudden demand for shagreen skins means that many skins come from fish of unknown origin, causing concern as to whether they have been farmed from sustainable sources, or if the product's sudden popularity risks decimating stingray populations. Far from being peripheral to the archaeologist's understanding of materials, such emotive debates highlight the complexity of the way people appropriate materials. This may be particularly true of leather, which often retains such a visual reminder of its animal origins.

Technological choices

As the modern materials science and engineering industry has often found to its cost, it is not enough to produce innovative materials with superlative properties and performance and expect them to be adopted for use; people are surprisingly dismissive of unfamiliar materials (National Research Council (US) 1999, ix-x). The means by which people accept and use materials is a complex and multifaceted web of properties, values, aesthetics and emotions. The appropriation of materials in a particular time and place is not only a matter of assessing the finished product, but also the technology of sourcing raw materials and techniques of production. This contextual interplay between production, properties, performance, distribution, appropriation and use is encapsulated in the theory of 'technological choices' (Lemonnier 1993; Sillar & Tite 2000, 6, fig.1). This approach recognises technology as a social construct, which is embedded in the perceptions and beliefs of local populations. It centres round the observation that there are many ways to achieve a technological goal, whether to spin thread, build a chariot or cover a table. The way which it is done, is a means by which archaeologists can understand that society.

An investigation of technological choices, achieved through the method of identifying the *chaîne opératoire* (operational sequence) is a way to question why certain technologies become the accepted approach. At first glance these may appear determined by material constraints while technical processes may be primarily seen as manipulating the material, but if one investigates a little deeper, material constraints emerge as just one aspect of the many contributing factors to how and why people go about producing and using whatever they do (Sillar & Tite 2000, 3). From this stance, we are encouraged to question the choices made at each stage of the *chaîne opératoire* in the production of an object. We may question; why this raw material, these tools, in this place, this crafts person, these techniques or this sequence of actions (Sillar & Tite 2000, 4)? Through being encouraged to answer these questions we can recognise the multiple influences on the maker and recognise not only a process of material transformation from skins to leather and leather to object, but that the maker is informed in these choices through his or her wider cultural beliefs and expectations. From Raistrick's leather sourced from the well cared-for Swiss cattle used to make textile machinery to spin thread for cloth, to Youngson Coll's opulent shagreen table top, to the leather chariot that drove a pharaoh into battle as described by Veldmeijer and Ikram, we see that the use of leather is deeply embedded not only in the capability of leather to perform these tasks, but also in the complex wants, ambitions and desires of people past and present and their ability to achieve them.

The authors and chapters

The authors in this volume address the question 'Why leather?' from their own specialist perspective. In the first two chapters, Amanda Michel and Eddie Cheshire present different approaches to identifying the material properties of specific leathers. In Michel's case, she uses her experience of microscopy in the contemporary leather industry to show how goat, sheep and cattle skins can be examined to predict how they behave in use. Cheshire uses his skills in mechanical and production engineering examining composite materials to test various methods of processing leather against the impact of ballistics. Through their different approaches these authors provide an understanding of the material properties of the raw materials for leather, then highlight the significance of processing skins into leather to create very different finished leathers. The next two authors, Laura Youngson Coll and Alan Raistrick are both directly involved with using leather to make products. Youngson Coll is a fine artist who is skilled in the traditional craft practice of bookbinding and leatherwork. Her chapter shows how a crafts person approaches leather commissions in a practical sense, as part of a collaborative design and craft team and then reflects on the role of such objects in contemporary society both in terms of ethics and status. Raistrick's chemistry background and family heritage in tanning, making leathers for the machinery used in the textile industry, show the great specificity and control over the tanning process and selection of raw materials that was required to gain exactly the right performance characteristics for each of the leather components.

The final two chapters by Barbara Wills working with Amanda Watts, and André Veldmeijer writing with Salima Ikram, provide the archaeological view. In these, the authors are working with archaeological evidence of leather objects, and building a comprehension of leather materials as a means to understand the role of leather in the societies under investigation. Wills and Watts are working predominantly with written and pictorial sources of wineskins in the Mediterranean, which are further elucidated using historical knowledge of the wineskin trade and the science of the contemporary wine industry. Veldmeijer and Ikram investigate the remarkably preserved leather remains of an ancient Egyptian chariot, and compare the dynamics of such a model with wooden examples as a way to understand the relationship of technique, materials and use. The eight authors possess a remarkable range of skills and perspectives, which together provide an innovative answer to the research question set for the conference. These feed into a wider current interest in archaeology to understand materials such as leather in the past.

Acknowledgements

With thanks to Roy Thomson and Ruth Whitehouse for their advice on an earlier draft of the introduction. Their knowledgeable comments sharpened the text, for which I am very grateful. I am indebted to the British Academy for providing financial support in the way of a Post-Doctoral Fellowship, for the project 'Cloth Cultures in Prehistoric Europe' held at the Institute of Archaeology, University College London. This project provided the time and intellectual foundation to develop this collaborative volume.

References

Angus, A. (2002). An introduction to the types of tannages use on ethnographic leather. In: Wright, M.M. (ed.). *The conservation of fur, feather and skin. Seminar organised by the conservators of ethnographic artefacts at the Museum of London on 11th December 2000.* London: Archetype Publications. 1-6.

Appleyard, H.M. (1978). *Guide to the identification of animal fibres.* Leeds: Wira.

Brandon-Cox, H. (1969). The trail of the Arctic Nomads. London: Kimber.

Charles, R. (1997). The exploitation of carnivores and other fur-bearing mammals during the North-Western European late Upper Palaeolithic and Mesolithic. *Oxford Journal of Archaeology,* **16**, (3). 253-277.

Collins, M., Buckley, M., Grundy, H.H., Thomas-Oates, J., Wilson, J., & Van Doorn, N. (2010). ZooMS: the collagen barcode and fingerprints. *Spectroscopy Europe,* **22**, (10) 6-10, available from: http://212-113-150-109.static.directrouter.co.uk/images/stories/ArticlePDfs/MS_22_2.pdf

Covington, A.D. (2006). The chemistry of tanning materials. In: Kite, M. & Thomson, R., (eds.). *Conservation of leather and related materials.* Oxford: Butterworth-Heinemann. 22-35.

Delong, M., Wu, J., & Park, J. (2012). Tactile response and shifting touch preference. *Textile: The Journal of Cloth & Culture,* **10**, (1). 44-59.

DeMarrais, E., Gosden, C., & Renfrew, C. (2004). Introduction. In: DeMarrais, E., Gosden, C. & Renfrew, C., (eds.) *Rethinking materiality: the engagement of mind with the material world.* Cambridge: McDonald Institute for Archaeological Research. 1-7.

Douglas, G.W. (1956). *Survey of the production of hides, skins and rough-tanned leathers in India, Pakistan, Ceylon and Africa.* Egham, Surrey: British Leather Manufacturers' Research Association.

Driel-Murray, C. van (2000). Leatherwork and skin products. In: Nicholson, P.T. & Shaw, I., (eds.). *Ancient Egyptian materials and technology,* Cambridge: Cambridge University Press. 299-319.

Driel-Murray, C. van (2002) Practical evaluation of a field test for the identification of ancient vegetable tanned leathers. *Journal of Archaeological Science,* **29**. 17-21.

Edwards, E., Gosden, C., & Phillips, R.B. (2006). Introduction. In: Edwards, E., Gosden, C. & Phillips, R.B., (eds.). *Sensible objects: colonialism, museums, and material culture.* Oxford: Berg. 1-31.

Falcão, L. & Araújo, M.E.M. (2011). Tannins characterisation in new and historic vegetable tanned leathers fibres by spot tests. *Journal of Cultural Heritage,* **12**, (2). 149-156

Graves-Brown, P. (2000). Introduction. In: Graves-Brown, P., (ed.). *Matter, materiality and modern culture,* London and New York: Routledge. 1-9.

Grey, T. (2006). *Archaeological Finds Procedures Manual.* With contributions by: Barham, L., Egan, G., Goodburn, D., Groves, J., Keys, L., MacConnoran, P., Maloney, C., Pearce, J., Smith, T. & Wardle, A., http://www.animalbones.org/Warwick%20teaching%20files/postex/MoLAArcFindsProcedures.pdf. MOLA, Museum of London. Museum of London Specialist Services. [Accessed: 08/03/2014.]

Groenman-Van Waateringe, W., Kilian, M., & Londen, H. van (1999). The curing of hides and skins in European prehistory. *Antiquity,* **73**. 884-890.

Haines, B.M. (2006a). Collagen: the leathermaking protein. In: Kite, M. & Thomson, R., (eds.). *Conservation of leather and related materials.* Oxford: Butterworth-Heinemann. 4-10.

Haines, B.M. (2006b). The fibre structure of leather. In: Kite, M. & Thomson, R., (eds.) *Conservation of leather and related materials.* Oxford: Butterworth-Heinemann. 11-21.

Harper, C. (2008). I found myself inside her fur... *Textile: The Journal of Cloth & Culture,* **6**, (3). 300-313.

Harris, S. (2014). Sensible dress: the sight, sound, smell and touch of Late Ertebølle Mesolithic cloth types. *Cambridge Archaeological Journal,* **24**, (1). 37-56.

Hodges, H. (1995). *Artifacts: an introduction to early materials and technology,* 2nd ed. London: Gerald Duckworth and Co.

Hollemeyer, K., Altmeyer, W., Heinzle, E., & Pitra, C. (2012). Matrix-assisted laser desorption/ionization time-of-flight mass spectrometry combined with multidimensional scaling, binary hierarchical cluster tree and selected diagnostic masses improves species identification of Neolithic keratin sequences from furs of the Tyrolean Iceman Oetzi. *Rapid Communications in Mass Spectrometry,* **26**, (16). 1735-1745 available from: DOI: 10.1002/rcm.6277

Hurcombe, L. (2007). A sense of materials and sensory perception in concepts of materiality. *World Archaeology,* **39**, (4). 532-545.

Ingold, T. (2007). Materials against materiality. *Archaeological Dialogues,* **4**, (1). 1-16.

Ingold, T. (2012). Toward an Ecology of Materials. *Annual Review of Anthropology,* **41**, (1). 427.

Jones, A. (2004). Archaeometry and materiality: materials-based analysis in theory and practice. *Archaeometry,* **46**, (3). 327-338.

Kellogg, K. (1984). *Home tanning and leathercraft simplified.* Charlotte, Vermont: Williamson Publishing Co.

Klokkernes, T. (2007). *Skin processing technology in eurasian reindeer cultures. A comparative study in material science of Sami and Evenk methods. Perspectives on deterioration and preservation of museum artefacts. PhD thesis.* Rudkøbing: LMR-Press.

Knappett, C. (2005). *Thinking through material culture: an interdisciplinary perspective.* Philadelphia: University of Pennsylvania Press.

Leather Conservation Centre (1981). *The fibre structure of leather.* London: The Leather Conservation Centre.

Lemonnier, P. (1993). Introduction. In: Lemonnier, P., (ed.). *Technological choices: transformation in material vultures since the Neolithic.* London: Routledge. 1-35.

MacGregor, G. (1999). Making sense of the past in the present: a sensory analysis of carved stone balls. *World Archaeology*, **31**. 258-272.

Mason, O.T. (1891). *Aboriginal skin-dressing - a study based on material in the U.S. National Museum.* Washington: Smithsonian Institute.

Meskell, L. (2005). Introduction: object orientations. In: Meskell, L., (ed.). *Archaeologies of materiality.* Oxford: Blackwell Publishing. 1-17.

Mould, Q., Carlisle, I., & Cameron, E. (2003). *Craft, industry, and everyday life: leather and leatherworking in Anglo-Scandinavian and Medieval York.* York: Council for British Archaeology.

National Research Council (US) Committee on Materials Science and Engineering: Forging Stronger Links to Users. (1999). *Materials science and engineering: forging stronger links to users / Committee on Materials Science and Engineering: Forging Stronger Links to Users, National Materials Advisory Board, Commission on Engineering and Technical Systems, National Research Council.* Washington, D.C.: National Academy Press.

Oakes, J. & Riewe, R. (1996). *Our boots: an Inuit women's art.* London: Thames and Hudson.

Ottaway, P. & Morris, C.A. (2003). The leatherworking tools recovered. In: Mould, Q., Carlisle, I. & Cameron, E., (eds.). *Craft, industry, and everyday life: leather and leatherworking in Anglo-Scandinavian and Medieval York, vol. 17: The small finds.* York: Council for British Archaeology. 3235-3244.

Paine, R. (1994). *Herds of the tundra: a portrait of Saami reindeer pastoralism.* Washington: Smithsonian Institution Press.

Rahme, L. & Hartman, D. (2001). *Leather; preparation and tanning by traditional methods.* Portland, Oregon: The Caber Press.

Raedler, C. (2007). Geräte aus Keramik in der spätbronzezeitlichen Ramsesstadt. Die Schaber der Werkstätten des Grabungsplatzes Q I. In: Pusch, E.B. & Bietak, M., (eds.). *Forschungen in der Ramses-Stadt. Band 5.* Hildesheim: Verlag Gebrüder Gerstenberg.

Reed, R. (1972). *Ancient skins, parchments and leathers.* London and New York: Seminar Press.

Richards, M. (2004). *Deerskins into buckskins: how to tan with brains, soap or eggs,* 2nd ed. Cave Junction, Oregon: Backcountry Publishing.

Rifkin, R.F. (2011). Assessing the efficacy of red ochre as a prehistoric hide-tanning ingredient. *Journal of African Archaeology*, **9**, (2). 131-158. Available from: http://www.academia.edu/2009297/Assessing_the_efficacy_of_red_ochre_as_a_prehistoric_hide-tanning_ingredient.

Schlumbaum, A., Campos, P.F., Volken, S., Volken, M., Hafner, A., & Schibler, J. (2010). Ancient DNA, a Neolithic legging from the Swiss Alps and the early history of goat. *Journal of Archaeological Science*, **37**, (6). 1247-1251.

Schwarz, S. (2002). Leathercraft in the mid-18th Dynasty. In: Andoin-Rouzeau, F. & Beyries, S., (eds.). *Le Travail du Cuir de la Préhistoire à Nos Jours. XXII rencontres internationales d'archéologie et d'histoire d'Antibes*. Antibes: Éditions APDCA. 481-493.

Sharphouse, J.H. (1983). *Leather technician's handbook.*, 23rd. rev. ed. Northampton: Leather Producers' Association.

Sillar, B. & Tite, M.S. (2000). The challenge of "technological choices" for materials science approaches in archaeology. *Archaeometry*, **42**, (1). 2-20.

Teerink, B.J. (1991). *Hair of West European mammals: atlas and identification key.* Cambridge: Cambridge University Press.

Thomson, R. (2006a). The nature and properties of leather. In: Kite, M. & Thomson, R., (eds.). *Conservation of leather and related materials.* Oxford: Butterworth-Heinemann. 1-3.

Thomson, R. (2006b). Testing leathers and related materials. In: Kite, M. & Thomson, R., (eds.). *Conservation of leather and related materials.* Oxford: Butterworth-Heinemann. 58-65.

Thomson, R. & Mould, Q., (eds.). (2011). *Leather tanneries: the archaeological evidence.* London: Archetype Press.

Tilley, C. (2007). Materiality in materials. *Archaeological Dialogues*, **4**, (1). 16-20

Toups, M.A., Kitchen, A., Light, J.E. & Reed, L. (2011). Origin of clothing lice indicates early clothing use by anatomically modern humans in Africa. *Molecular Biology and Evolution*, **28**, (1). 29-32.

Wilder, E. (1976). *Secrets of Eskimo skin sewing.* Anchorage, Alaska: Alaska Northwest Publishing Company.

Wildman, A.B. (1954). *The microscopy of animal textile fibres: including methods for the complete analysis of fibre blends.* Leeds: Wool Industries Research Association.

Skin deep: an outline of the structure of different skins and how it influences behaviour in use

A practitioners' guide

Amanda Michel

Introduction

Leather can be made from the skin of virtually any vertebrate animal: mammal, reptile, bird or even fish. The key to leather's unique properties is the protein collagen that comprises the major component of skin of all animals.

Collagen

Proteins come in many forms. There are some that are soluble in water, for example, the protein albumen that is found in egg whites. These are known as globular proteins. Then there are proteins that are insoluble in water, like the fibrous proteins found in meat. Collagen is one of these fibrous proteins and is the protein in skin that ultimately becomes leather. Collagen is the most abundant of the animal proteins: more than a quarter of the body's protein is collagen. It is not only found in skin, it also occurs in large quantities in tendon, cartilage and bone. Fresh, unprocessed hide or skin contains around 33% protein, of which 29% is collagen. There are believed to be as many as nineteen different types of collagen that have different functions within the body (Bailey & Paul 1998, 104). In skin, the most abundant types are I and III with type IV forming links between the epidermis and dermis that ultimately becomes the grain surface of the leather (Bailey & Paul 1998, 104).

It is the intricate structure of collagen that enables leather to be produced, which is incredibly strong and durable yet highly flexible. Looking at it at a molecular level, collagen is made of thin strands of amino acids, largely glycine, proline and hydroxyproline, joined together to form a poly-peptide chain. The arrangement of the amino acids is unusually regular in collagen with glycine taking up every third space in the chain (Covington 2009). Three of these chains twist together to form a helical collagen molecule that is only 1.5 nanometres in diameter. Several collagen molecules twist together to form micro-fibrils. Microfibrils twist together to form fibrils (typically 100 nanometres in diameter). Fibrils group together to form fibres (one micron in diameter) and then the fibres group together to form fibre bundles. This continuous twisting of ever increasing strands of collagen has a structure that is very similar to the way rope is made (figure 1). This is why collagen has similar properties to a rope; it is very flexible, yet very strong.

Bundles of collagen fibres then interweave with one another in a three-dimensional manner, which further increases strength. It is these interweaving collagen fibres that create the unique fibre structure of skin that ultimately becomes leather (figure 2). Up to 4% of the proteins found in skin are the non-structural globular proteins. It is important that these are removed during the leather making process otherwise they stick the collagen fibres together after tanning thus making the leather firm and inflexible (Sharphouse 1983, 23).

Leather fibre structure in cross section

If leather is observed in cross section, it can be seen that the fibre structure changes through its thickness (see Haines 2006).

The grain

At the top is the smooth surface that would have been covered with hair before the tanner processed it and used chemicals to remove it. This is called the grain surface. Below the surface there is the *grain layer* that contains many different structures (figure 3):

Figure 1. Rope. iStock.

Figure 2. Photomicrograph of the cross section of bovine leather showing interweaving fibre bundles. Photographed at X50 magnification. Photograph by Amanda Michel.

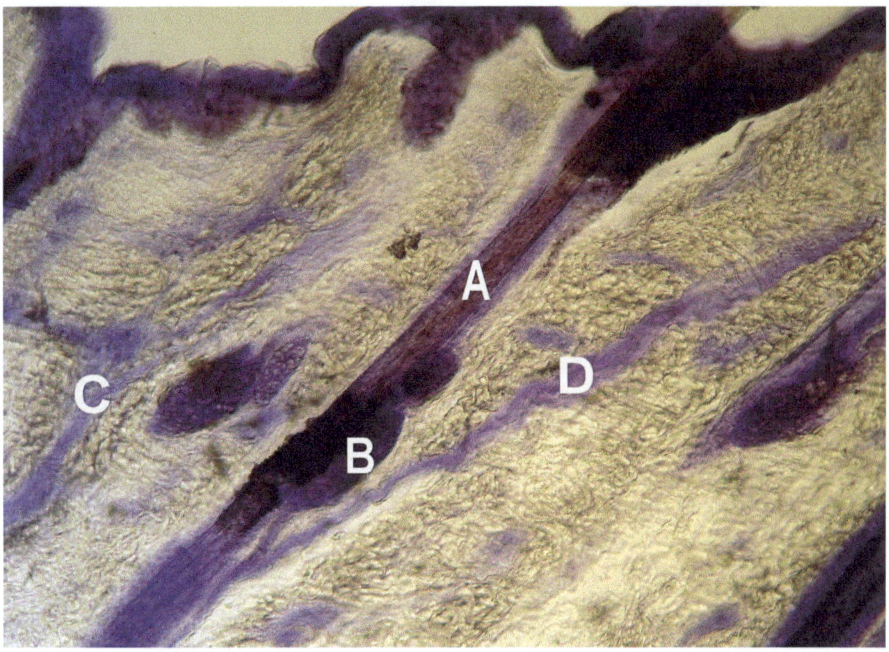

Figure 3. Photomicrograph of the cross section of the grain layer of cattle hide showing different structures. Photographed at X100 magnification. A = hair follicle; B = sebaceous gland; C = blood vessel; D = erector pili muscle. Structures revelead by the histological stain Azure A". Photograph by Amanda Michel.

- *Hair follicles* from which the hair grew (A);
- *Sebaceous glands* that coat the hair with greasy substances to protect it from the weather (B);
- *Blood capillaries* that bring blood to the surface of the skin which also helps to regulate body temperature and supply the other structures with nutrients (C);
- *Erector pili* muscle that raises the hair (D);
- *Sweat glands* that produce perspiration, which helps regulate body temperature.

Due to the many structures that occur in the grain layer, there is limited space available for the formation of large collagen fibre bundles; only a limited number of much smaller fibres form. Since it is collagen fibres that give leather its strength, the grain layer on its own is relatively weak.

The corium and flesh layers

Below the grain layer is the corium layer. This part of the skin comprises mostly collagen fibre bundles and therefore the corium imparts most of the strength to the leather. Research has shown that the corium is up to *ten* times stronger than the grain layer (O'Leary 1996) (figure 4). At the bottom there is the flesh layer. This part of the skin contains fat and connective tissue and was once in contact with the muscle beneath. Since it is of little use for making leather, it is usually removed early on in the leather making process.

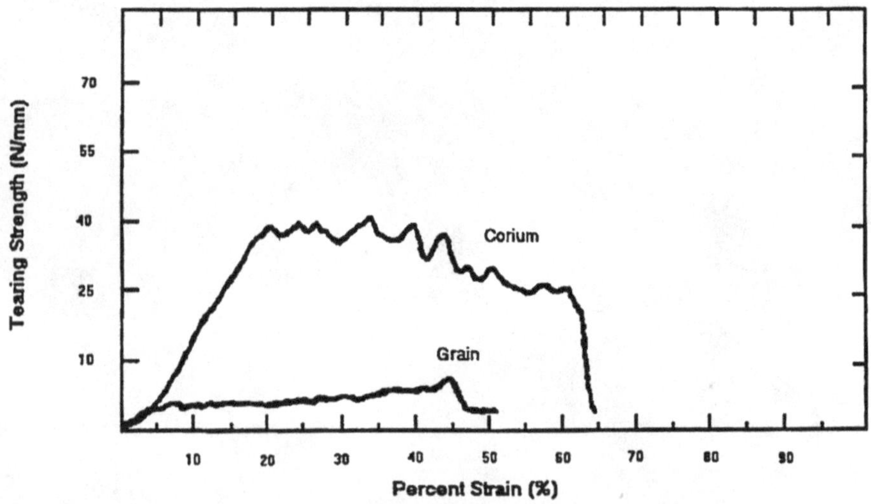

Figure 4. Graph showing the difference in tear strength between the grain and corium layers of leather. Courtesy of the University of Northampton.

Leather fibre structure – areas of the hide

Not only does the structure of leather change throughout its thickness, it also changes across the skin. Being of natural origins, the structure of the leather very much relates to the way the animal lives. The skin can be roughly divided up into three distinct areas: the neck and shoulder, the belly and the butt (figure 5).

Neck and shoulder

The skin is very thick in the neck and shoulder area. This is because in the wild, animals are attacked by predators which generally attack at the neck. The skin therefore has grown thicker here for extra protection. The neck area also has many wrinkles which are a consequence of the constant bending of the neck as the animal grazes (figure 6).

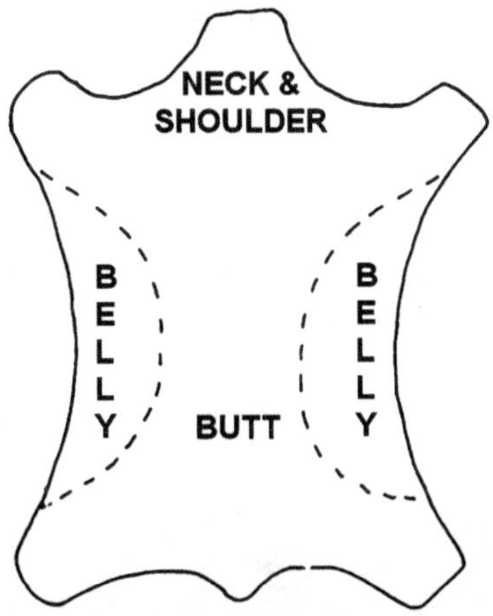

Figure 5. Diagram showing the different areas of the hide. By Amanda Michel.

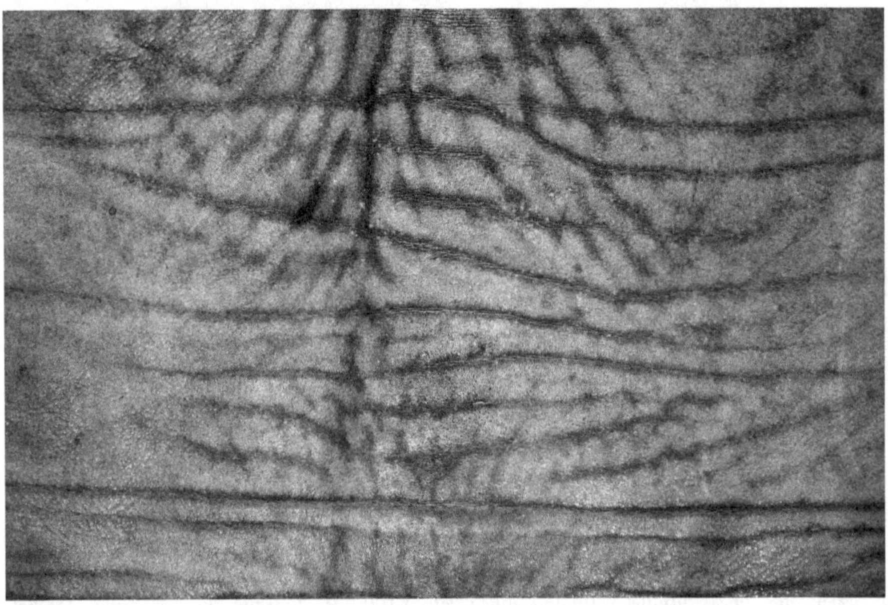

Figure 6. Wrinkles in the neck area of leather. Photograph by Amanda Michel.

Belly

The skin is very thin and stretchy in the belly area because, in life, the belly has to be able to expand and contract a lot, for example, when the animal eats. The collagen fibres in the belly area have to be able to move over one another to enable the skin to stretch easily. This is made possible by larger spaces between the fibres. Consequently, the fibres are not compacted together as much as they are in other areas (figures 7 & 8). This phenomenon can often be easily seen if the back of the leather is viewed: the belly area has a more fibrous appearance than other areas due to the larger spaces between them (figure 9). Because of the inherent stretchiness of the leather in the belly, it is not suitable for use in areas such as seat panels in upholstery or knees/elbows in garments because it will stay in a semi-permanent stretched state and appear baggy. However, it is perfectly suitable for making, for example, the sides and back of the furniture or collar and pocket flaps of a jacket.

Axillae

At the edge of the bellies are four small pockets that have a much looser and more fibrous appearance than the rest of the belly. These are called the axillae and are where the limbs join the body (figure 10). The leather here is very thin and stretchier than elsewhere: in the living animal this area of the skin had to be very flexible to allow free movement of the limbs. The axillae do not react to wear very well at all; they stretch and become baggy very quickly.

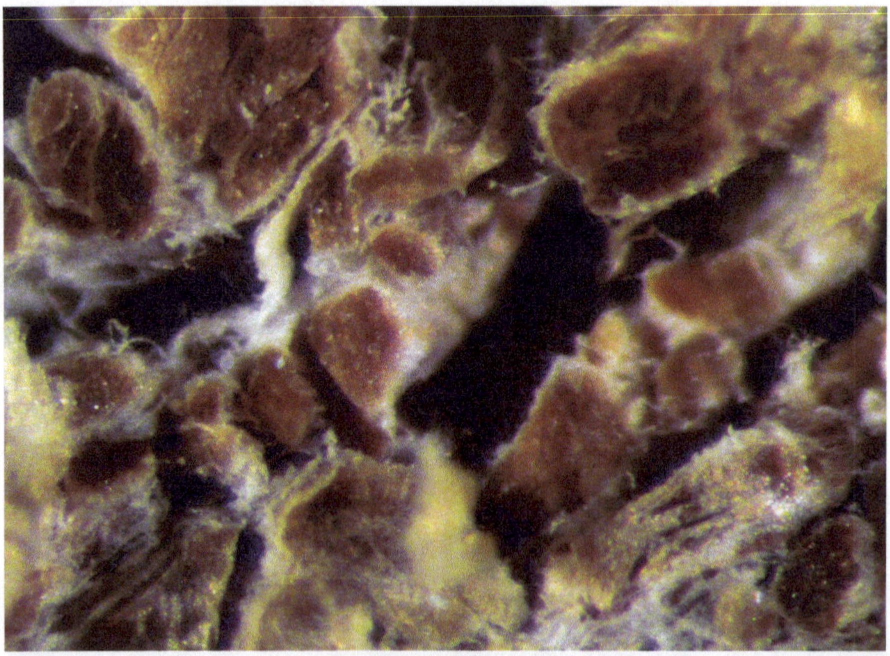

Figure 7. Photomicrograph of the cross section of the belly area of leather showing spaces between the fibre bundles. Photographed at X100. Photograph by Amanda Michel.

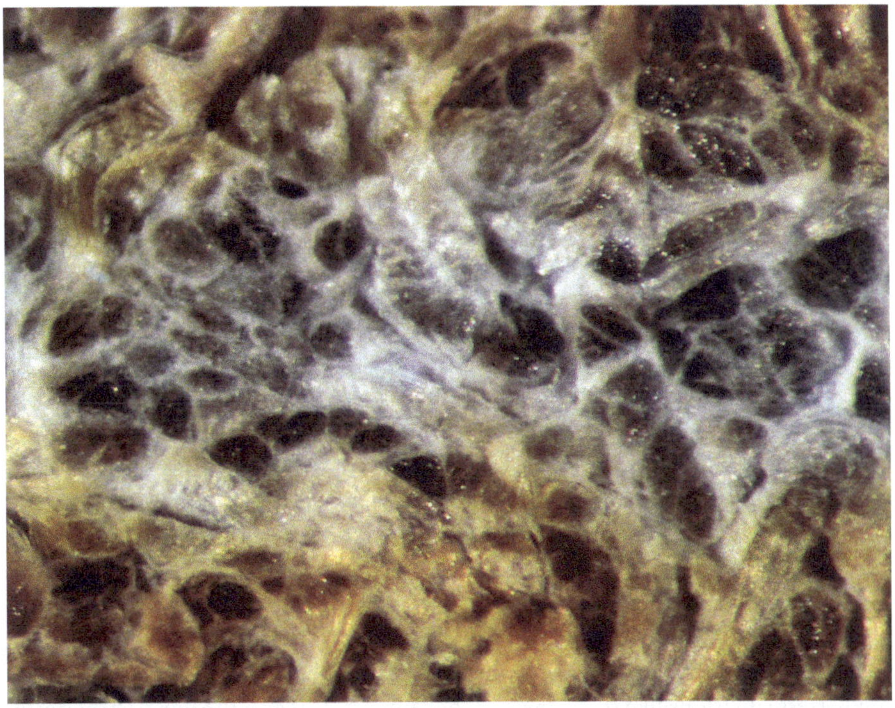

Figure 8. Photomicrograph of the cross section of the butt area of leather showing a more compact fibre structure. Photographed at X100. Photograph by Amanda Michel.

Figure 9. View of the back of leather showing a more fibrous appearance in the belly area. Belly = left; butt = right. Photograph by Amanda Michel.

Figure 10. View of the back of leather showing a much more fibrous appearance in the axilla area. Photograph by Amanda Michel.

Butt

The area that once covered the back and rump of the animal, known as 'the butt', has the most uniform and compact structure of all. Therefore, the butt is the strongest and most durable area of the skin. So, this is the prime cutting area for high-stress situations like seat panels or shoe uppers. It is also the most aesthetically pleasing to the eye as it will withstand wear better and not become as wrinkled as leather made from the other areas does.

Although the thickness of any skin varies considerably in different locations, the overall thickness can also be influenced by the age and sex of the animal. Generally, male animals tend to have thicker skins than female and this difference increases with age.

Leather fibre structure - stretch

The fibres in hides and skins are not randomly orientated; they run in a direction that closely follows the direction of hair growth. If leather is pulled in the same direction as the collagen fibres it will be much more stretchy than if pulled in the opposite direction (figure 11). When cutting leather for use, it is important to consider this variation in stretch when placing the patterns. For example, footwear needs to be able to stretch in certain directions to enable it to be lasted properly. Therefore, the patterns must be orientated to allow the stretch to take place over the shoe and not along the shoe whilst still maximising cutting yield; the patterns are placed with the heel to toe in the direction where there is less stretch. This is often called 'tight to toe' in the footwear industry.

Variation in structure and grain pattern between species

As previously mentioned, leather can be made from the skin of any vertebrate animal. The majority of the leather, however, is made from the skins of domestic animals: cow, sheep, goat and pig. Skins from hunted animals are used, such

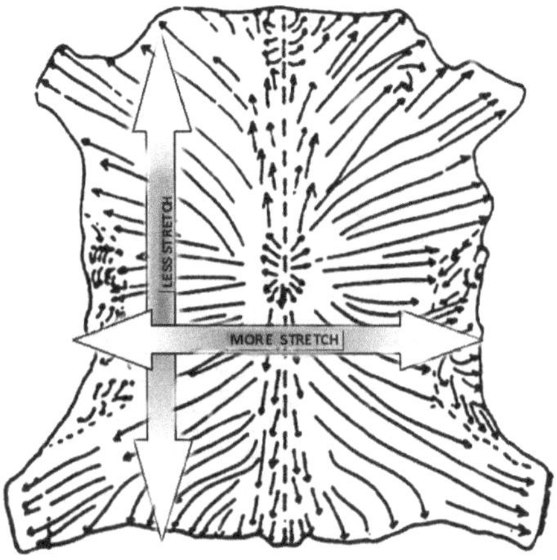

Figure 11. Diagram of the directional growth of the collagen fibres and the resulting difference in stretch. Original diagram courtesy of the University of Northampton, Amanda Michel.

as deer skin, but, compared to domestic animals, form a small part of the total leather produced today. Some skins might be suitable for some purposes but not for others. This may be due to something as fundamental as size: for example, it is not possible to cut large sofa panels out of a small sheepskin, but there are other reasons too. The differences in structure of different hides and skins can have a profound effect on the leather's performance in use.

The hair that once covered the skin, before its conversion into leather, grows from tubular structures known as 'hair follicles'. These tubes open out at the surface of the leather and the pattern that they make is known as the 'grain pattern'. Each species usually has its own characteristic grain pattern that can be used to distinguish leather from different species.

Cattle

The grain pattern of cattle hide leather has follicles which are all of similar size and are fairly evenly distributed (figure 12). Full thickness adult cattle hide can be a centimetre thick or more in some areas. Clearly this is much too thick for most purposes other than soling shoes or for some industrial uses. Therefore, the leather usually is split into layers of more manageable thickness.

In cattle hide leather that has not been split into layers, the grain layer accounts for approximately one fifth of the total thickness of the hide, depending on the location (figure 13). Remembering that the grain layer on its own is relatively weak, when leather is split to the required thickness for use, it is essential that sufficient corium is left to give the leather enough strength. Therefore, to make very thin leather the correct hides must be selected. Although the thickness of the hide depends on the age and sex of the animal, the ratio of approximately one part grain to four parts corium remains the same whatever the hide thickness.

To make thin leather, it is advisable to select a hide that has a thin grain layer such as a calf skin, thus ensuring more of the strength-giving corium will be present when the leather is split into layers (figure 14). But, calf leather tends to be more expensive and has more limited cutting area due to its smaller size. If a

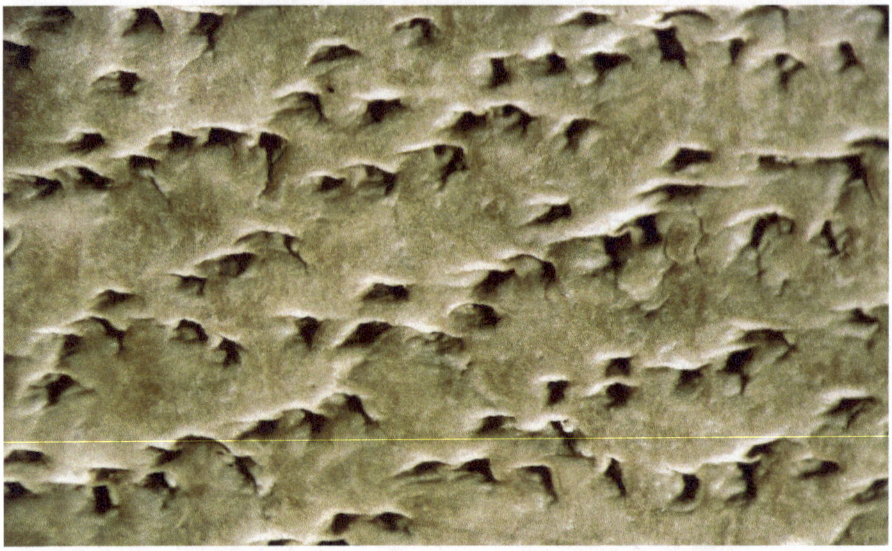

Figure 12. Grain surface of cattle hide leather. Photographed at X25 magnification. Photograph by Amanda Michel.

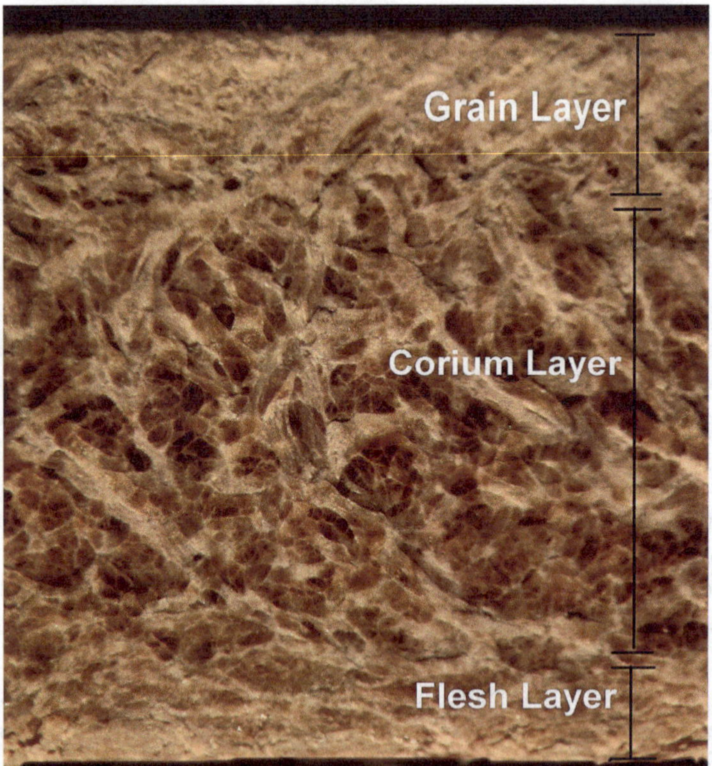

Figure 13. Cross section of full thickness cattle hide leather showing the different layers and the ratio of grain to corium. Photographed at X25 magnification. Photograph by Amanda Michel.

Figure 14. Cross sections of calf leather (left) and adult cattle leather (right) showing differences in grain layer depth resulting in more corium layer remaining after splitting to the desired thickness. Grain layer depth indicated in yellow. Splitting level indicated in red. Photographed at X25 magnification. Photograph by Amanda Michel.

larger cutting area is required, the use of cow hide (female) may be the answer since they are generally thinner than steer hides (male). Using cow rather than steer gives both thinness and size as cows are usually slaughtered when they have finished their productive milking life and are several years old. Hides from male beef animals are smaller because they are slaughtered at a much younger age (12 to 18 months old). However, the belly area of cow hides tends to be stretchier than the belly area of male (steer) hides, so attention must be paid to where certain panels are cut from. Because bovine leather is strong and durable and can be made into leathers of varying thicknesses, it has many applications from soft clothing leather to hard shoe sole leather.

Sheep

Sheepskins can be processed with the wool still on, in which case they are known as woolskins. However, the majority of sheepskins have the wool removed and are processed into what is known as grain leather. Because they have been selectively bred to produce large quantities of wool, the grain pattern of sheep leather has more hair follicles per unit area than that of most other animals. The hair follicles are, compared to cattle hide leather, relatively small and closely packed together in groups with spaces between the groups (figure 15).

In cross section sheepskin leather is different from cattle hide. Being much thinner than cattle hide leather, it is not normally necessary to split sheepskin. Since almost half the total thickness of sheepskin is grain layer, weakness would result if a significant amount of the existing corium was removed (figure 16). The collagen fibre bundles in

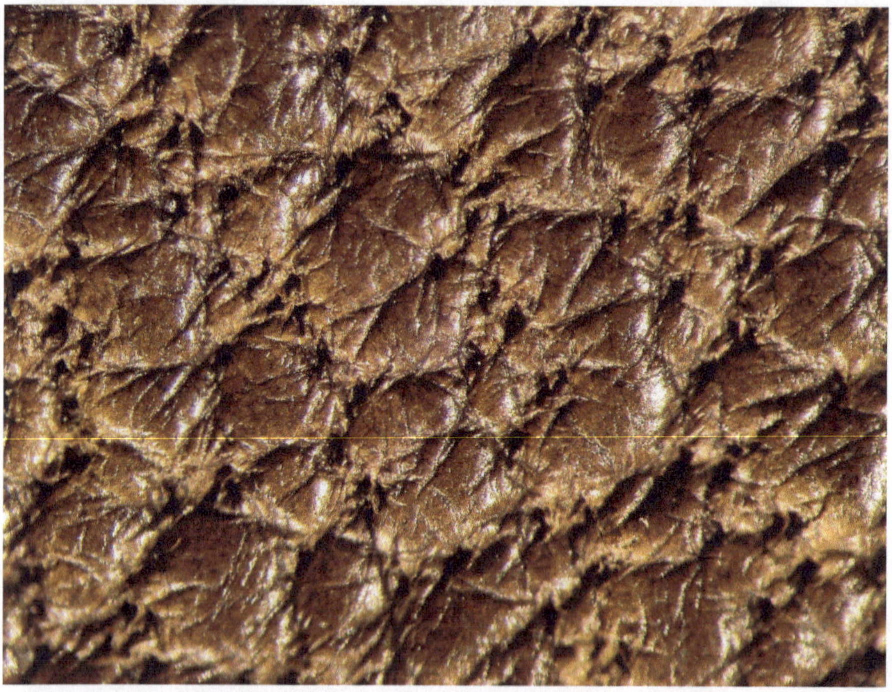

Figure 15. Grain surface of sheepskin leather. Photographed at X25 magnification. Photograph by Amanda Michel.

Figure 16. Cross section of full thickness sheepskin leathe. Photographed at X25 magnification. Photograph by Amanda Michel.

the corium are much smaller than those of cattle hide. This means that sheepskin can be used to make suede with a fine, velvety nap that is favoured for clothing purposes.

The much denser structure of the grain layer in sheepskin, results in their skins having lots of hair follicles. This results in more sebaceous glands, blood vessels and related elements. The presence of these structures in the grain layer reduces the available space for the strength giving collagen fibres and thus the grain layer of sheepskin is weaker than that of many other leathers.

Unlike the gradual transition from grain to corium found in cattle hide leather, in domestic sheepskin the structure suddenly changes and sometimes there is a gap between the grain and corium (figure 17). This is a result of sheep's unusual characteristic of a layer of fat between the grain and corium; most other animals deposit fat between the skin and muscle tissue below (subcutaneous fat). The tanner has to remove this fat with detergents to enable the tanning chemicals to penetrate properly. If the skin is very fatty, after it has been degreased there can be gaps left where the fat used to be. When this happens the grain layer is able to move on top of the corium to the point where it can become detached in places. This phenomenon is sometimes called 'looseness' or 'delamination'. The more the leather is flexed, the more the layers delaminate. The tanner has to take great care during processing to prevent this happening. But it is not always possible to stop it happening once the leather has been made up into a product. This is one of the reasons why sheepskin is not suited to situations where vigorous flexing is likely to take place, such as footwear or seat cushions. Sheepskin leather is most commonly used in clothing.

Figure 17. Cross section of sheepskin leather showing delamination between the grain and corium layers. Photographed at X25 magnification. Photograph by Amanda Michel.

Goat

The grain pattern of goat leather is very characteristic. There are usually two different sizes of hair follicle that are arranged in lines; the line of large ones is closely followed by a line of small ones (figure 18). This is because goats often have two types of hair in their coats; short, fine hairs called 'secondary hairs' (other names for secondary hairs include 'underfur' and 'undercoat'), and longer and stiffer outer hairs called 'primary hairs' (also referred to as 'guard hairs', 'outer hairs', or 'outer coat').

Goats are of a similar size to sheep and therefore their leather cannot be used where a large cutting area is required. Unlike sheepskin, the fibre structure of goat leather is more like that of cattle hide. As goats are not wool-producing animals, the density of hair follicles in the grain layer is similar to cattle hide and therefore the grain layer is slightly stronger than that of sheep. Also, goats do not deposit fat in the middle of the skin like sheep, so there is more of a gradual transition from grain to corium (figure 19). The result is a lightweight skin that can withstand flexing more readily than sheepskin. It is therefore more versatile and can be used for a wide variety of products such as garments, lightweight footwear, gloving and some smaller items of furniture. Like sheepskin, the collagen fibre bundles are quite small enabling a very fine 'suede' to be produced.

Figure 18. Grain surface of goatskin leather. Photographed at X25 magnification. Photograph by Amanda Michel.

Figure 19. Cross section of full thickness goatskin leather. Photographed at X25 magnification. Photograph by Amanda Michel.

Hair sheep

There are animals known as 'hair sheep' that look very similar to goats and have a skin structure that is more similar to goat than the domestic wool sheep. However, although they are very goat-like, genetically they are sheep. Hair sheep skins are very popular for gloving leather as they have a fine grain pattern and are quite stretchy. They originate predominantly from African countries.

Pig

The grain surface of pigskin has a very coarse, nodular appearance. The skin grows like this for protection purposes; because the skin is not protected with lots of fur like other animals, the skin grows these nodules to minimise the trauma to the skin of the animal caused by scratches, etc. (figure 20). This feature is passed on to the leather. Pigskin leather is very durable and abrasion resistant which makes it ideal for uses where this is important, such as saddle seats. Depending on the location on the skin, the hair follicles of pigskin are often arranged in a triangle shape. The fibre structure of pigskin is quite different to that of the other commonly used skins.

Instead of having a grain layer that contains the hair follicles and a corium layer below, the hair follicles in pigskin penetrate right through the full thickness of the leather; they can be seen from both surfaces. So, technically, pigskin is all grain layer with no corium (figure 21). Therefore, one might expect the leather to be very weak. But this is not the case as pigskin leather is very strong, because there are few hair follicles present; pigs do not have a thick fur coat like cattle or sheep, they just have a sparse covering of coarse bristles. Consequently, there is plenty of space between the hair follicles for strength giving collagen fibres. The drawback

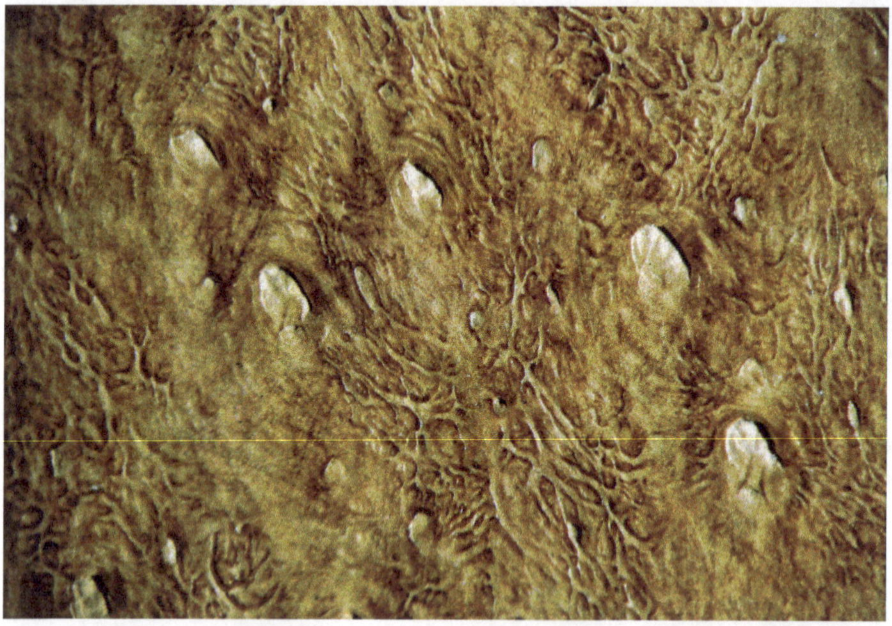

Figure 20. Grain surface of pigskin leather. Photographed at X25 magnification. Photograph by Amanda Michel.

Figure 21. Cross section of full thickness pigskin leather showing that the hair follicles penetrate through the full thickness of the skin. Photographed at X25 magnification. Photograph by Amanda Michel.

to this lack of a protective fur coat is their skin gets scratched very easily, resulting in poor quality leather.

Since the follicles penetrate right through the leather and form holes in the leather, it is not possible to make a water resistant leather from pigskin. The hair follicles can always be seen from both sides of the leather. Sometimes they are particularly prevalent on the suede side of the leather; a problem called 'fish eye' (figure 22). In some areas of the skin, the collagen fibres that surround the hair follicle are particularly fine. As they have a larger surface area, they tend to dye a different colour and appear as a pale spot in the suede. In the middle is a very small black dot where the hole of the hair follicle is, hence the reference to fish eyes. The tanner can minimise this effect to a certain extent by careful processing and choice of dye, but it is an inevitable characteristic. The collagen fibres are quite small in pigskin, so the suede has a fine nap.

Since there is no concern about causing weakness by cutting away too much of the corium, pigskin can be split relatively thinly whilst still retaining strength. This enables the leather to have a variety of uses from hardwearing saddlery to soft clothing suede. It is also particularly suited for use where very thin leather is required such as bookbinding and small leather goods.

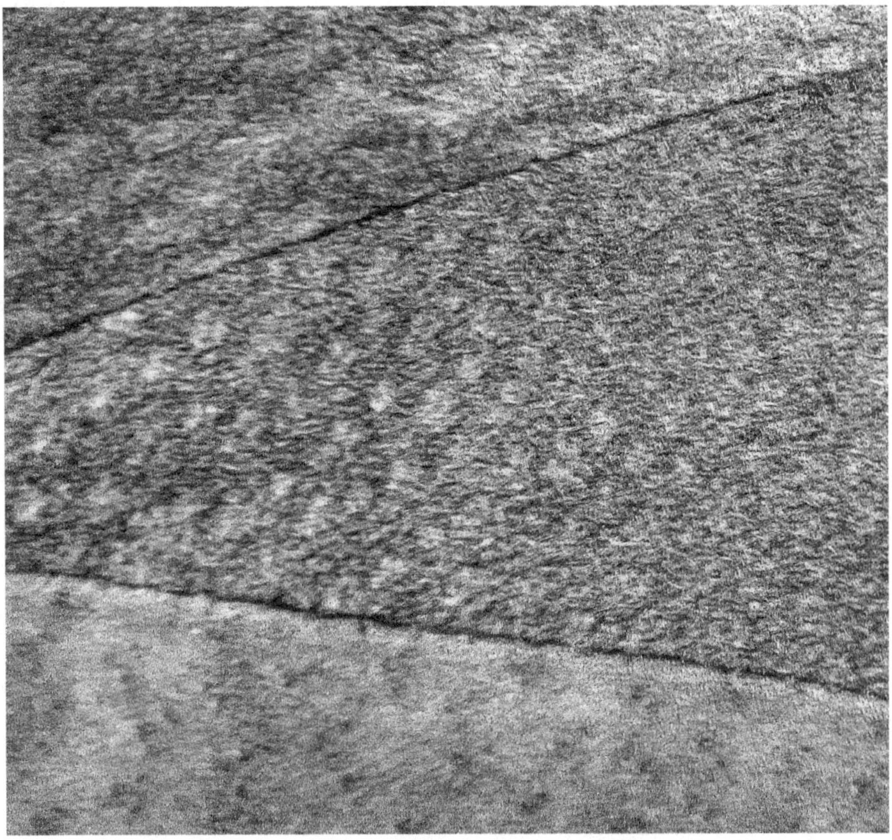

Figure 22. Suede surface of pigskin leather showing fish eye effect. Photograph by Amanda Michel.

Conclusions

The aim of this paper is to provide a better understanding of why leather continues to be the preferred material for a wide variety of products due to its behaviour in use. The examination of leather structure by cross section, area, stretch and species provides insight into why the skins from certain animals are preferred for certain products and not others. It is also hoped that the simplified manner in which the information is provided will enable those less experienced with leather to be more able accurately identify the species origins of leather in artefacts based on the grain pattern and internal structure of the leather.

References

Bailey, A.J. & Paul, R.G. (1998). Collagen: a not so simple protein. *Journal of the Society of Leather Technologists and Chemists,* **82**, 3. 104-110.

Covington, A.D. (2009). *Tanning chemistry, the science of leather.* Cambridge: Royal Society of Chemistry.

Haines, B.M. (2006). The fibre structure of leather. In: Kite, M. & Thomson, R. (eds.). Conservation of leather and related materials. Oxford: Butterworth- Heineman. 11-21.

O'Leary, D. (1996). *Differences in strength between the grain and corium layers of bovine leather.* Unpublished PhD Thesis, University of Northampton.

Sharphouse, J.H. (1983). *Leather technicians' handbook.* Northampton: Leather Producers' Association, rev. ed. 23.

Further Reading

British Leather Manufacturers' Research Association (1957). *Hides, skins and leather under the microscope.* Egham, Surrey: BLMRA.

Cuir bouilli armour

Edward Cheshire

Introduction

In Medieval times the material known as *cuir bouilli* was widely used for armour, but its precise composition has puzzled historians. Insofar as any consensus has been reached, it has generally been regarded as leather which had been hardened either by immersion in heated water or by impregnation with wax (Blair 1979, 19; Singer 1957, 171). This paper reviews evidence on the subject derived from literature and the chemistry of leather and hide, and sets up a series of experiments into the manufacture of non-metallic armour and the defensive properties of *cuir bouilli* against arrows, with a main purpose of eliminating from the discussion those semi-mythical materials which are demonstrably not up to the job and assessing what remains.

The experiments use merit indices to rank the efficacy of various armour materials, based on the ratio of Depth of Wound/Areal Density (the Areal Density of armour is the mass per unit area; with units of kg per square metre). The experiment was carried out in two stages. The first results were obtained by shooting arrows at bovine rawhide and leather, which had been prepared by different methods involving heat and water. It was found that boiled rawhide performs well against arrows, with a merit index at least six times better than leather, whether natural, boiled or wax impregnated. The merit index is a numerical value calculated to rank different properties in a simple, readily understood order.

The second stage of the experiment tested shooting arrows at hard-faced, hide based armour. Such material was advocated by an Arab writing at the time of the Crusades (Nicolle 2002, 203-204). The results show that this technique, employing a coating mainly of crushed rock in hide glue, was found to improve the merit index by a further multiple of three, and also to increase the tendency for arrows to ricochet. The wearer of such armour could expect to emerge from combat either unscathed or to receive an arrow wound typically only one fifteenth deep, or less, as that sustained by someone putting their trust in, for example, waxed leather: say 10 mm or less deep compared with a lethal 150 mm or more (Cheshire 2010, 239-294).

Background and literature survey

Mention of *cuir bouilli* is not uncommon in accounts from the Middle Ages. An early instance occurs in the *Chanson d'Antioche* in 1185 (Foulkes 1912, 100). The earliest usage in English noted in the Oxford English Dictionary was in 1375 by Barbour, followed by Chaucer in about 1386 (Simpson & Weiner 1989). The term, in a variety of spellings, was employed for over a century and is recorded in the same dictionary as last being used during that era by Douglas in 1513. It re-entered the lexicon in about 1880, but it should not be expected that the sense was identical after more than three centuries in the etymological wilderness.

Any convincing argument about the nature of *cuir bouilli* must reconcile the demonstrable properties of reproductions of it with the Medieval literature, scant though the latter is. The name itself, from Norman French, is generally translated as 'boiled leather'. However, in Norman French *'cuir'* had a much wider meaning than just leather, embracing also skin, hide, rawhide and even flesh (Hindley 2000, 171; Kelham 1978, 198-200). The interpretation as 'boiled leather' produces difficulties for such authors as Waterer and Cameron, who correctly noted that although boiling leather does yield a substance that is hard, it is also brittle, and therefore useless as armour (Cameron 2000, 25-33; Waterer 1981, 62-77). They have, together with Dobson (2003, 79-99), argued that the substance was leather which was not boiled but treated in some other fashion. Waterer proposed that wet leather, after being moulded by a process known as 'samming', was heated to no more than 50^0 C, whereas Dobson reconstructed a pair of vambraces (leather armbraces), from wet leather dried in an oven at 70^0 C. Cameron suggested that *cuir bouilli* might have been made by basting leather with hot wax and rosin. Apart from the name itself, the only reference known to the author which gives a clue about the manufacture of *cuir bouilli* armour comes from the chronicler Froissart, who mentions it when describing the siege of a town on the coast of north Africa during 1390. Reportedly the Saracens covered their targes with "*cuir bouilly de Cappadoce, ou nul fer ne peut prendre n'attacher, si le cuir n'est trop* échaufé" (Froissart 1874, 471). The present author translates this as "cuir bouilly of Cappadoce (Cappadocia in modern Turkey), which no iron is able to attach (penetrate), if the hide is not too heated." It seems that heat was involved in the manufacturing process, but that overheating might cause damage. Since Froissart makes a point of mentioning this potential problem it may not have been unusual (see below, 'The effect on hide of water and heat').

No evidence has been found from contemporaneous sources that making *cuir bouilli* armour involved impregnation with waxes, oils, rosin or any other foreign substance. These are recent introductions to the discussion. Confusion also emanates from those leather workers who nowadays dip moulded leather briefly in hot water to make jugs and similar items, which they call '*cuir bouilli*' (Anfield 1982, 128-129). It is proposed here that '*cuir*' was rawhide and '*bouilli*' was the past participle of the verb '*bolir*', to boil (Hindley 2000, 171) or '*boviller*', also to boil (Kelham 1978, 32).

An avenue of armour research consistent with the evidence, but apparently unexplored until recently, is the boiling of rawhide in water and hide glue. The technique was certainly used with success by Native Americans to make shields during the 19th century:

> *"...the Sioux shield made of the skin of the buffalo's neck, hardened with the glue extracted from the hoofs and joints of the same animal. The process of* 'smoking the shield', *is a very curious, as well as an important one in their estimation. For this purpose a young man about to construct him a shield, digs a hole of two feet in depth, in the ground, and as large in diameter as he designs to make his shield. In this he builds a fire, and over it, a few inches higher than the ground, he stretches the rawhide horizontally over the fire, with little pegs driven through holes made near the edges of the skin. This skin is at first, twice as large as the required shield; but having got his particular and best friends (who are invited on the occasion), into a ring, to dance and sing around it, and solicit the Great Spirit to instil into it the power to protect him harmless against his enemies, he spreads over it the glue, which is rubbed and dried in, as the skin is heated; and a second busily drives other and other pegs, inside of those in the ground, as they are gradually giving way and being pulled up by the contraction of the skin. By this curious process, which is most dexterously done, the skin is kept tight whilst it contracts to one-half of its size, taking up the glue and increasing in thickness until it is rendered as thick and hard as required (and his friends have pleaded long enough to make it arrow, and almost ball proof), when the dance ceases and the fire is put out" (Catlin 1973, 241).*

The effectiveness of Native American shields in combat has been remarked upon by Bill, a surgeon in the United States army. His comments, at the end of a long article on the treatment of arrow wounds, were made at Fort Craig, New Mexico, in 1862 (365-387):

> *"We wish in conclusion to recommend to those in authority the plan of protecting soldiers and others exposed to arrow wounds with a light cuirass. The Indians have a method of dressing bulls' hide for shields for themselves, which renders it arrow proof. A cuirass made of such material, protecting the whole trunk before and behind, need not weigh more than eight or ten pounds, and by means of it a soldier could enter an Indian fight with a fair chance of escaping death [...] We are certain that if suitable bulls' hide cuirasses were provided fatal wounds from arrows would become very rare."*

Regrettably, no evidence has been found that surgeon Bill's suggestion, made some 600 years after such cuirasses were worn in battle in Europe, was taken up. The word '*cuirass*' itself has a coriaceous derivation, from the Old French '*cuirace*' (from the Latin corium), meaning 'dermis' or 'layer of skin' (Tulloch 1992, 348).

One technique to improve the resistance of armour to penetration is to harden the impact face. Perhaps the earliest existing detailed description of face hardened armour is that of al-Tarsūsī, who wrote in Egypt for Saladin in the late 12th century AD. Several comprehensive instructions for its manufacture are quoted by Nicolle (2008, 153-163) and a relevant passage is given here:

"The scrapings of rawhide/leather are well soaked in water. Then they are pulverised and are mixed with half as much shavings of wood and of cobbler's gum/glue and half of [dried?] spleen or blood well ground. Then take some moulds of clay or wood in the shape which you require for the plates [lamellae or hoops] of the cuirass. Spread these on a table and cover each with a piece of rag or skin. Take a small amount of filings of a hard metal, some shavings of rawhide/leather, some gut and some powdered horn. Put all this in a pot, throw in about half as much of fish glue and 'black glue', and heat until it is softened. Then throw in about a third [as much] of the first mixture; grind it fine in a stone mortar and pulverise it in a grinder until it is well blended and perfectly mixed. Smear it across the moulds which you have covered with skin and rag. Put in sufficient, then let it dry; smear with spleen, and sprinkle over it filings of shāburqān *['hard iron' or 'natural steel']. Afterwards file all of them, polish them, equalize then (sic) [for shape and size], and varnish them. If you wish you can encrust them with ivory and pierce them with holes [for lacing?]. While they are still supple, place some wooden pegs across[ways?] and, during one dry time [day], pierce them with a drill and 'arrange them' [...] Of the same mixture you can create a helmet by changing the mould and going through the same process already described, after which you can embellish, varnish, enrich, gild, varnish with 'varnish of* farāgh*' ['empty' or 'finishing', perhaps meaning clear varnish], in three 'hands' [meaning unclear], depending on what you want." (Between [ ] inserted by Nicolle).*

This armour has three distinct layers. Starting at the rear face, the 'skin' or 'rag' contributes long fibre reinforcement which, once impregnated with a setting liquid, becomes a tough, high tensile composite material. The second layer combines short fibre reinforcement, from the wood, shavings of rawhide/leather, gut and powdered horn, with hard metal filler. The material has some strength and toughness, although less than the backing, combined with an increased hardness. The final layer contains only hard metal filings in a resinous matrix, covered with varnish to complete the encapsulation of any partially wetted particles and provide protection from water. This surface will be hard but brittle. This Medieval process nicely illustrates the basic principle of face-hardened armour: a graduation from a hard but brittle impact face through to a tough, high tensile strength backing.

The impact faces of some of the several types of armour described by al-Tarsūsī, are heavily filled with particles of such materials as glass (typically mostly SiO_2), marble (a hard crystalline metamorphic rock resulting from the recrystallization of a limestone, $CaCO_3$), emery (aluminium oxide, Al_2O_3) and 'hard iron' (probably iron with a high carbon content). The first three of these are mineral fillers, characterised by high hardness with high compressive strength and modulus, but low toughness.

Al-Tarsūsī's filled mixes have various glues as the matrix, such as fish glue, spleen and 'black glue'. All of these will have a much lower Young's Modulus than the fillers and probably have a superior toughness. The Young's Modulus is the ratio of the stress to the strain in a material. A material with a high Young's Modulus is relatively 'stiff' or 'rigid'. However, when a material contains a high proportion of filler, "as much as will be soaked up", the particles are either very close to each other, or even touching. Any tensile strain in the material will have relatively little effect on the stiff fillers, but cause a disproportionately high strain in the low modulus matrix, with associated cracking and failure.

The chemistry and porosity of skin

The strength of skin, treated or otherwise, comes mainly from collagen, which comprises the largest proportion. The chemistry is complex, but a simple model is helpful to an understanding of how *cuir bouilli* may have been made. References in this paper to 'leather' relate only to a vegetable tanned product.

Collagen

Collagen is an insoluble fibrous protein, comprising long chain molecules built up from amino acids (see also Michel, this volume, for a summary of the chemistry of skin). These are covalently linked through the acidic portion of one amino acid and the amine portion of another. Each of the 18 different amino acid types in collagen can form the basic amide link, analogous to the junction between two links in a steel chain. Each acid also has additional atoms forming a side cluster, known as an 'R-group', which can be visualised as projecting sideways from each link. These R-groups vary considerably in size, and some are so large that they interfere with each other and distort the way the chain hangs.

A fibre of collagen consists of three protein chains twisted round each other in a helical manner, similar to the well known DNA molecule (which has only two strands). The three chains are not free floating but are connected at frequent intervals, mainly by hydrogen bonds between a hydrogen atom in one chain and an oxygen atom in a neighbouring chain. The R-groups are located on the outside of the helix and able to protect it to some extent from outside interference (figure 1).

Hydrogen bonds are not as strong as the covalent bonds and under certain conditions are liable to be attacked by water molecules which, also consisting of hydrogen and oxygen, interpose themselves between the original hydrogen and oxygen atoms in the protein chains. When sufficient hydrogen bonds have been attacked, the helical structure of collagen collapses and the fibre becomes much shorter and fatter. For collapse to occur two basic conditions are necessary. First, a sufficiently high temperature, known as the shrinkage temperature (Ts), must be reached. Secondly, sufficient water must be available to penetrate between the R-groups round the outside of the helix and into the core, where the hydrogen bonds are.

Nature ensures that skin has a shrinkage temperature high enough to withstand the environment in which the animal lives, plus a handsome safety margin. Thus Ts of the skin from cod, a deep cold water fish, is 40^0 C, whereas that of bovine skin is about 65^0 C (Gustavson 1960, 279-304). At the level of our simple model this can be explained by the different proportions of amino acids in the skins, with some R-groups more effectively hindering water from reaching the hydrogen bonds in the core of bovine collagen.

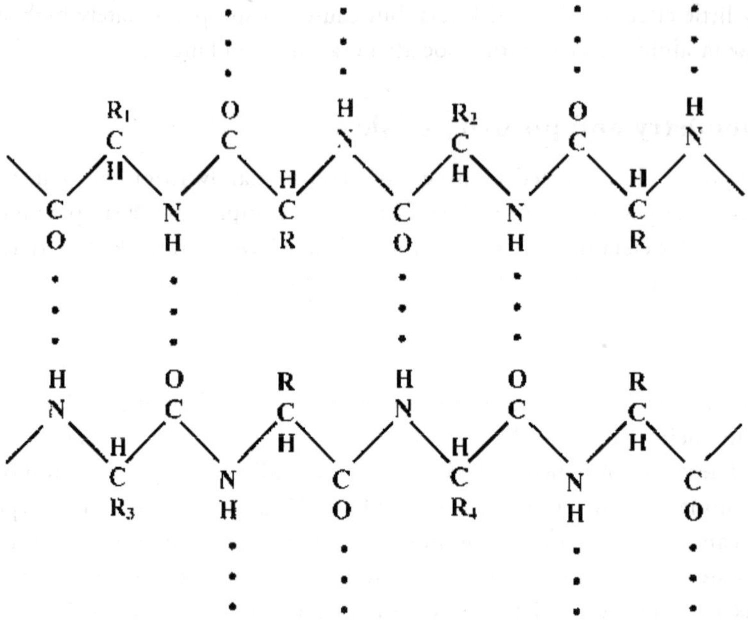

Figure 1. Hydrogen bonds (dotted lines) between protein chains in collagen. The bonds at the top and bottom of the diagram loop back to link with the third protein chain (not shown) to form a helix, analogous to three strands in a rope. Diagram by Edward Cheshire.

The shrinkage temperature is generally quoted for skin which is thoroughly wet, so that an adequate quantity of water is readily available to attack the hydrogen bonds. However, it has been observed that the shrinkage temperature increases as the proportion of water in the skin falls. Eventually, a point is reached, for vacuum dried skin, where shrinkage does not occur. For cowhide this is at about 145^0 C, which is the same as that reported for collagen by Witnauer (1960, 441-451).

The solid line in figure 2 shows the variation in the shrinkage temperature for unmodified hide containing 25 to 62% by weight of water, with an extrapolation to 145^0 C for dry hide. It can be seen that material containing less than about 22% of water will not shrink at a temperature of 100^0 C. It follows that dried rawhide will not shrink substantially in boiling water unless and until it has absorbed more than 22% of the liquid. Witnauer (1960, 441-451) also observed that the maximum amount of water that is effective in lowering the 'melting point' of bovine collagen is about 62%. With this degree of saturation the material shrinks at 56^0 C.

Figure 2 also shows vegetable tanned leather (dotted line). The shrinkage temperature for this material when containing large amounts of water is slightly higher than that for hide. This is because a primary purpose of a tanning agent is to react chemically with the collagen and to reduce the vulnerability of the protein chains, with their critical hydrogen bonds, to attack by water. Tanning also creates additional hydrogen bonds (Ramanathan & Nayudamma 1960, 455). On the other hand, leather containing small amounts of moisture shrinks at a lower

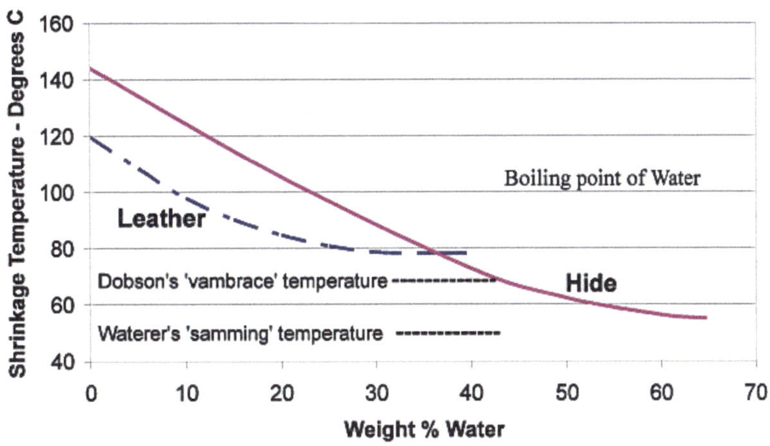

Figure 2. Graph of shrinkage temperature against water content. From: Witnauer, 1960: 441-151.

temperature than rawhide: the graph shows that leather containing more than about 10% of water will shrink at 100^0 C.

Prolonged boiling in water will break down the protein chains in collagen into gelatin, a transparent water soluble protein. This can be produced by boiling pieces of hide in water and evaporating the liquor, leaving gelatin as a hard residue. This substance can readily be re-dissolved in hot water and employed as hide glue. Gelatin can also be used in food preparation, so armour based on hide has the unique advantage that it can, in extremis, provide some nutrition. Josephus (1959, 340) narrates of the Jews besieged in Jerusalem by the Romans:

> *"In the end they actually devoured belts and shoes, and stripped off the leather from their shields and chewed it."*

The 1559 expedition by Tristran de Luna to Carolina also fell on such hard times that they were forced to eat the leather of their shields and the trappings of their horses (Chard 1940, 94).

Porosity

A further property of skin-based material, which has relevance to the production of *cuir bouilli,* is its porosity. When freshly flayed hide is allowed to dry it shrinks significantly in all directions, and the volume therefore reduces. If the skin has been restrained at points on the perimeter, the resulting rawhide will have a characteristic scalloped edge. The effect, or something very similar, can be seen on the greaves in some Medieval illustrations. This material is semi-transparent, demonstrating that there are within the structure few inclusions, such as air bubbles, to scatter the light and hinder its transmission.

By contrast, when hide is tanned into leather the overall dimensions are substantially unchanged. However, the process removes the gelatinous material between the collagen fibres and only partially replaces it with chemically reacted vegetable tanning agent. The detail of the structure of the resulting leathers depends on the processing conditions, which are selected by the tanner to suit the end use, but the voids make the material porous to a considerable extent (figure 3).

If a piece of leather is plunged into water it will quickly saturate, but rawhide will only absorb water at a very much slower rate. This is very important to the aspiring manufacturer of *cuir bouilli*.

The effect on hide of water and heat

It was observed that when a piece of vegetable tanned leather is plunged into boiling water it promptly floods. The water content of the collagen rapidly increases to much more than the critical 10% by weight at which leather will shrink at 100^0 C (figure 2). The effect is an almost instantaneous contraction of the leather and the colour changes to black. The material becomes plastic and can readily be moulded to shape, but work must be speedy because once removed from the hot water it rapidly cools and becomes very hard and rigid, confirming comments by Lingwood (1990, 64-71).

The properties of boiled leather can be explained in terms of the model introduced above. The helical structure of collagen in leather or hide is maintained principally by the hydrogen bonds, which are attacked by boiling water. The three protein molecules in each helix are no longer mutually supporting and collapse to a fraction of their original length. The leather contracts in two dimensions but also increases in thickness. When the material cools new hydrogen bonds form, but these are not in the original highly ordered positions. The new bonds occur

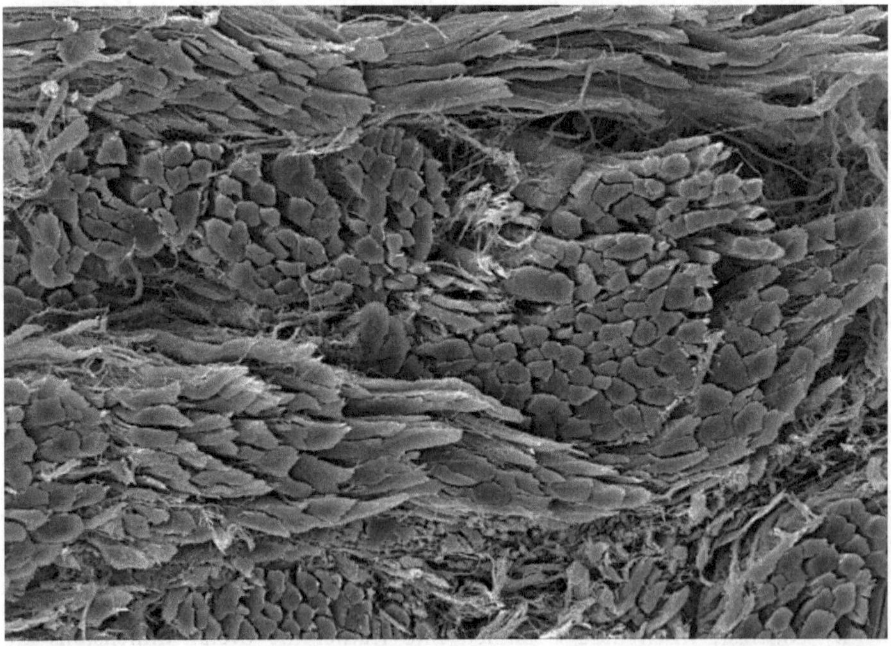

Figure 3. Section through vegetable tanned leather showing fibrous structure and the voids which cause porosity. Magnification X500.

between oxygen and hydrogen atoms which happen to be in close, but random, proximity as the temperature falls. The final product no longer has the aligned helical molecular form that formerly gave it strength and toughness. It is hard but brittle and the very poor resistance to impact renders it useless as armour.

Figure 2 also throws light on the 'samming' method suggested by Waterer (1981), who limited the temperature to 50⁰ C. This is well below the shrinkage temperature of even fully wet leather, so very few molecular bonds would be broken and reformed. Any hardening would be insufficient, as was observed by Cameron (2000, 31) (some bonds in fully saturated leather may break at a lower temperature than the neat line on the graph suggests, perhaps because some of the many different amino acid residues are more susceptible than others. The tanning might also be incomplete).

Dobson (2003) deliberately employs leather tanned at the two surfaces but not in the interior, so called 'scabbard butts'. He flirts with the shrinkage temperature of leather, and would achieve more hydrogen bond breakage than Waterer. However, working at 70⁰ to 80⁰ C, he says that he "can tell that the leather has "cooked" enough when it is trying to contract" (Dobson 2003, 89). Figure 2 suggests that this incipient contraction may be attributed, at least partly, to the un-tanned hide in the centre of the scabbard butts exceeding its shrinkage temperature. Rawhide in boiling water increases in thickness but reduces in area, as does leather (figure 4).

However, the effect on rawhide is very much slower and can be interrupted at any time. The edges are the first to swell, because water penetrates at the cut sides as well as through the faces. The edges also reduce in length and the material tends to dish with the flesh surface on the concave side. This face is first to shrink because the fibrous structure is more open here, allowing easier access by water. Eventually, the whole piece becomes a roughly uniform increased thickness, although it continues to absorb water indefinitely (figure 5). For armour, it is preferable to terminate the process after no longer than about 8 to 15 minutes, depending on the hide thickness, when the material has hardened but not lost much toughness. The thickness, hardness and areal density all tend to increase as the toughness, and hence the resistance to cracking, reduces. A briefer immersion

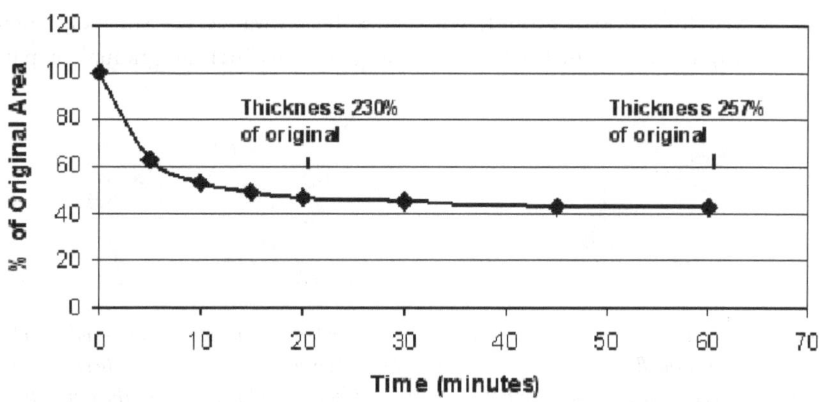

Figure 4. Areal shrinkage of 4.4 mm thick rawhide in boiling water. Diagram by Edward Cheshire.

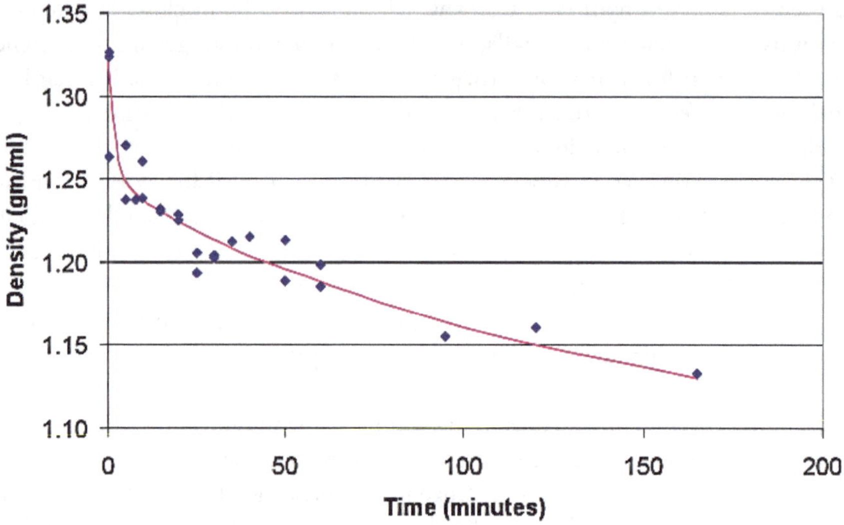

Figure 5. Change in density of rawhide immersed in boiling water. Diagram by Edward Cheshire.

can suffice for thin components, or when some resilience is required. The shape of the final item is also an important consideration, because if it is not restrained the rawhide initially curls and then may partially flatten again. Considerable skill and experience is necessary if a craftsman is to make allowance for both shrinkage and distortion without compromising the mechanical properties.

The moisture content of rawhide, and hence the shrinkage, can be controlled by preconditioning with damp rags. This may afford uniformity throughout the material, or introduce more water at either, or both, surfaces (with experience the worker can judge the dampness of rawhide from the flaccidity). The ingress of water can also be controlled by painting one or both sides of the rawhide with a waterproof substance. After immersion the material tends to curl away from a painted face (figure 6).

A rawhide 'preform' of a piece of armour may have been produced by restraining freshly flayed hide over a mould, with suitable allowance for shrinkage before and after any boiling. Still further processing options include gradual or partial

Figure 6. Rawhide after immersion in boiling water. The left side is natural material which has curled toward the flesh side because this face has taken up water faster. The flesh side on the right has been painted to slow down the ingress of liquid, so the curl is in the opposite direction. The cut edges swell first because of the larger area exposed. Photograph by Edward Cheshire.

immersion in boiling water or steaming in a box, and an empirical appreciation of figure 2, in conjunction with an equally empirical appreciation of Fick's laws of diffusion and thermal conduction (for an explanation of Fick's law, see Newey & Weaver 1990, 188-193), would have placed a very wide range of possible techniques within reach of Medieval armourers.

Coating *cuir bouilli* armour with a drying oil, such as linseed, would limit the ingress of water and reduce the tendency to rot, especially in a humid climate. The lacquer used to coat Japanese rawhide armour, and the tung oil on some Tibetan armour, would have performed a similar function (LaRocca 2000, 113-132; Kōzan 1962, 59-60).

Returning briefly to Froissart's '*échaufé*' remark, the author's experiments have demonstrated that boiling rawhide for a period of more than about half an hour in water does indeed cause progressive deterioration of the product. The water becomes brown in colour, indicating that material is leaching out. The rawhide turns rubbery and is much easier to cut and mould than after a shorter immersion. After a considerable period in air it eventually dries with a hard surface but the damage has been done; there is not only redistribution of the hydrogen bonds but also hydration of the amide links causing fragmentation of the protein chains. The mechanical effect on leather and rawhide of boiling is discussed below (see 'Mechanical testing').

Medieval armourers may well have been tempted to overboil *cuir bouilli* to a point of easier workability. Froissart's comment suggests that the practice was not uncommon and perhaps giving the material a poor reputation. The Native Americans producing shields (see above) apparently avoided this potential pitfall by applying only glue. This would make available less water for hydration and their technology would retain any leachate in the hide, although it would permit the manufacture only of near flat shapes (some of the more successful armour samples in the trials were rawhide boiled in a solution of hide glue in water).

The production of specimens of '*cuir bouilli*' and other materials

The techniques employed for this research were deliberately rudimentary to give confidence that they were not beyond the reach of a Medieval armourer. Even inexperienced craftsmen should, with a little practice, be able to make a far better item of armour than was managed by the author. After shrinkage the specimens measured only about 100 by 75 mm and were intended to be representative of the panels incorporated into a brigandine, as well as larger pieces of armour (Robinson 2002, 79-80).

The tools used were knives, saws and kitchen vessels, with clamps to counter the tendency of rawhide and leather to curl when heated in water. The author used gas for convenience but, regardless of the heat source, the temperature of boiling water (100°C at sea-level) would of course be the same through the ages. Experience would probably enable the craftsman to produce the best armour, but a repeatable time of hide immersion could be obtained from an hourglass or water leaking from a container if required. The leather was bovine and vegetable tanned in all cases. The rawhide was also bovine.

Some eighty specimens of various materials were employed in the shooting trials and a summary of the production methods is given below. For the purposes of comparison, some non-metallic materials other than hide based were included in the trials.

I) Boiled leather

Leather panels were typically immersed for one or two minutes in boiling water. They promptly turned black and curled up. They were then with some difficulty placed between large jaws in a vice and pressed approximately flat until cold and dry. In a few early attempts the water had salt (NaCl) or hide glue (gelatine) dispersed in it as a solution or gel. However, these experiments were not pursued because it rapidly became clear that the properties of boiled leather were such that it had little value as armour. The tensile strength was less than 5 N/mm^2 and it was so brittle that a blow with a hammer shattered it;

II) Boiled rawhide

The basic technique was the same as for boiled leather (including hide glue) but the immersion times were much longer. The optimum balance of properties was obtained with typically 8 to 12 minutes of immersion;

III) Waxed leather

Leather was immersed until saturated in bee's wax melted in a vessel. It was then removed and allowed to cool on a flat surface (the melting point of bee's wax is 63^0 C);

IV) Ram's horn

After the tips of rams's horns were cut off they were cut down one side. Heating them by fire (blowlamp) softened them until they could be pressed flat between the jaws of a vice. Work must be speedy before the material cools down;

V) Impregnated linen

Linen fabric was saturated with a strong concentration of hide glue in boiling water. Several layers were then placed on a flat surface until cold and dry. This technique produces a fibre reinforced composite material with sufficient mechanical properties to significantly reduce wounding by arrows;

VI) Soft armour

The soft armour in the trials consisted of multiple layers of linen, cotton and silk fabric, usually backed by cotton wool padding;

VII) Hard faced boiled rawhide

Once the rawhide-based product had dried (see II above), a hard surface coating of mainly mineral filler in hide glue was applied where appropriate. The manufacturing techniques were simple, employing only kitchen utensils such as pots, to heat the hide glue matrix, and knives to spread the mixtures, which cooled and dried to give a hard surface. This was again in part to ensure that Medieval technology would have been up to the job (figures 7 & 8).

The surface coating was applied only to some of the boiled rawhide. It could also have been used with profit on the impregnated linen, but there was no supporting evidence found in the literature that this material was used historically. Impregnated linen was included because it would be a straightforward way to make a 'muscled *cuirass*' in a clay or plaster mould and the performance against arrows showed it offers reasonable protection. If it was not used historically by the ancient Greeks (who did use linen armour), they certainly missed a trick (Everson 2004, 110-114; 145-152).

Mechanical testing

Mechanical test equipments were used to compare the relative impact resistance and tensile strength of various armour specimens.

Charpy impact testing

Figure 9 shows the Charpy test equipment. The pendulum swings down to strike the specimen, which has been cut to a standard size and shape, and the amount of energy absorbed by the impact is shown on the dial. The normalised fracture energy is given in Joules. The absolute values have little value here but the relative values are useful for comparison. The relative energies absorbed are shown in table 1 (no sample was coated).

Figure 7. A typical specimen of hard-surfaced boiled rawhide. In this case the surface coating comprises carborundum grit and finely crushed marble in a matrix of hide glue. The scratches are from an arrow ricocheting without penetrating. Photograph by Edward Cheshire.

Figure 8. A side view of a specimen of boiled rawhide with a surface coating of crushed marble in hide glue. The arrowhead projects 9 mm from the rear face. Photograph by Edward Cheshire.

Figure 9. Charpy Impact tester. Photograph by Edward Cheshire.

Rawhide	25J
Ram's Horn - Transverse	4J
Rams' Horn – Exterior on impact face.	6J
Linen Fabric - Impregnated with hide glue	5J
Boiled Leather	<1J
Boiled Rawhide (28 min)	<1J
Boiled Rawhide (14 min)	<2J
Boiled thick Rawhide (time not recorded)	7J

Table 1. Relative energy required to fracture various materials in Charpy Impact Test.

The boiling of rawhide caused a large reduction in fracture energy, which became worse the longer the material was treated. After about half an hour of boiling, rawhide deteriorated to have similar impact properties to briefly boiled leather.

The Charpy results indicated a more severe deterioration of boiled rawhide than seems consistent with other mechanical tests, or the shooting trials. It was observed that rawhide boiled in hide glue typically had a brittle surface layer of glue. This would promote cracks to run through the material and cause failure without absorbing much energy.

Tensile testing

A standard Instrom Test Machine was used for tensile testing of 'dog bone' shaped test samples of hide etc. Strain measurements were taken with a Demec mechanical gauge.

Figure 10 shows the drastic effect on vegetable tanned leather of being boiled in water. In this sample the tensile strength was reduced from some 23 N/mm^2 to 3 N/mm^2. Furthermore, the strain at fracture was much reduced, *i.e.* the material had become far more brittle.

Compare figure 10 with figure 11, which shows the stress/strain graphs of rawhide, both boiled and un-boiled. Rawhide deteriorates when boiled, but the change is very much slower than with leather. Even after 38 minutes, rawhide retains over 25% of its original strength and exhibits a yield point, *i.e.* it is less brittle than boiled leather. The plot for the latter is barely visible on the graph, near the origin (Specimen 08-22).

Hardness testing

A Barcol Portable Hardness Tester was employed to compare the hardness of the various specimens. The apparatus works by driving a pointed impressor into the surface with calibrated force. The model of tester available was not best suited to the hardness found, so a zero reading was given by the softer materials (as a rough datum from common experience, a well cured polyester resin such as those widely used to produce fibreglass boat hulls should give a Barcol reading of 40 to 50).

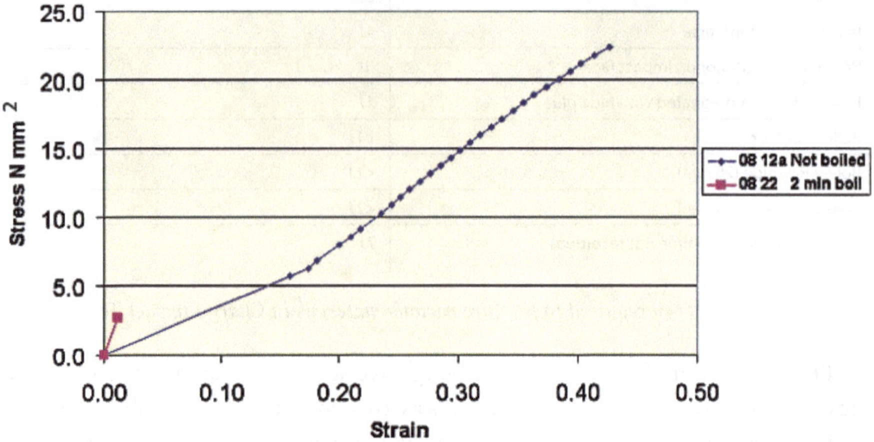

Figure 10. Sample Stress-Strain Graphs of leather (08-12a) and boiled leather (08-22).

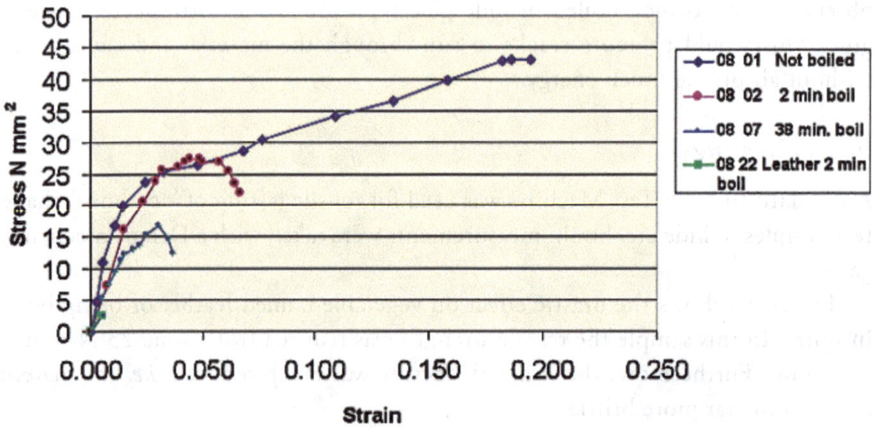

Figure 11. Sample Stress-Strain Graphs of natural rawhide (08-01), boiled rawhide (08-02 and 08-07) and boiled leather (08-22).

The absolute values in table 2 are of limited value but they do provide a hierarchy of relative hardness. In general the values are probably too low because the tester should be used on dead-flat surfaces, which was not the case with many samples.

Fracture toughness testing

An attempt was made to measure fracture toughness using a theory based on work by Griffith into the effect of flaws such as cracks in materials (Weidman 1990, 313). The testing involved the tensioning to failure of specimens with cracks running in from the sides but in practice these 'cracks', which should in theory have zero width, had to be cut with a bandsaw and were about one mm wide. The stress at the tip of the 'crack' was therefore much reduced compared with a true crack.

Material	Barcol hardness
Leather	0
Leather impregnated with paraffin wax	0
Woven linen impregnated with hide glue	0
Rawhide	0
Boiled leather - external face	5 - 10
Boiled leather - internal face	25 - 30
Ram's horn - external face	5 - 15
Ram's horn - internal face	25 - 35
Boiled rawhide	15 - 35
Boiled rawhide faced with marble in hide glue	35 - 40
Boiled rawhide faced with marble and slate in hide glue	30 - 40
Boiled rawhide faced with marble and carborundum	30 - 40
Cured Unsaturated Polyester Resin (reference only)	40 - 50

Table 2. Barcol hardness of a selection of armour specimens.

Further complication stemmed from the way in which rawhide initially hardened from the outside surfaces when boiled. After a short boil the rawhide in effect became two different materials, with the relatively tough original core sandwiched between *cuir bouilli*. Rawhide, boiled or otherwise, has a highly fibrous structure and may be regarded as a composite material with a high elongation to break, which further reduces the validity of 'crack' theory. Also the strain rate resulting from an arrow strike is many times greater than that imposed by a tensile testing machine. The Charpy tests gave more valid comparisons in this respect at least.

For the several reasons outlined above the fracture toughness measurements were judged to be of such complexity and doubtful accuracy that they will not be considered further here.

Principles of arrow penetration

Arrows were selected as the best weapons for comparative tests on the various armours. Historically, any combatant on a Medieval battlefield might have found himself exposed to arrow shooting, and it is also relatively straightforward to reproduce consistently in the laboratory.

The equipment for shooting arrows shown in figure 12 comprised an air gun with a barrel of 1½ inch calibre (38 mm) and a length of 6 ft (1.82 m). A typical arrow is shown in figure 13. The flights have been replaced by a light piston seal and a nylon spider at the middle of the shaft keeps the projectile, which has the dimensions of a crossbow bolt rather than an arrow, in the centre of the bore. The shaft is ash and the arrow heads mild steel. The projectile was in free flight at impact. To avoid, or at least minimize, damage to the arrows, the armour system was backed by several centimeters of plasticine. The mass of the test arrows was about 70 grams and the impact velocity was nominally 36 m per second. These values are based on published information with an allowance for aerodynamic

Figure 12. A view of the air gun looking from the breech end towards the arrow containment box, the lid of which is open. The larger red tube in the foreground is the air pressure vessel and the butterfly valve to release the compressed air to propel the arrow is yellow. The gun is muzzle loaded. Photograph by Edward Cheshire.

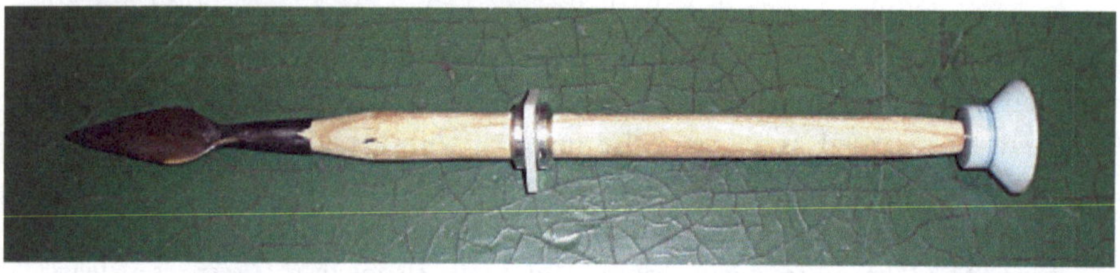

Figure 13. Typical trials arrow with piston seal and central 'spider'. Photograph by Edward Cheshire.

drag retardation out to a range of about 150 m (Bergman *et al.* 1988, 658-670; Richardson 1999, 50-53).

The target was hit with a kinetic energy of typically 40 joules. This is potentially lethal, although lower than the energy attributed to the longbow of the late 14th and 15th centuries AD, by which time *cuir bouilli* was becoming obsolete.[1] An important part of the armour system was the padding placed immediately behind the armour to reproduce an aketon (a padded garment worn under armour). High speed photography showed this being compressed under the impact and then rebounding vigorously. The significance of this is that the average force required to stop a projectile is inversely proportional to the stopping distance (work = force x distance). Thus, for example, the average force required to stop an arrow in 30 mm is only one tenth of that which will stop it in 3 mm.

The performance of body armour involves a complex interaction between the projectile, the armour itself, the backing padding and the human body. Some parts of the latter are better able to withstand deflection or penetration than others;

1 The often quoted 80 joules required to cause a casualty comes from German work in about 1900 with soft lead shrapnel balls, which are not pointed (Courtney-Green 1991, 13).

compare a buttock with a kneecap. A further advantage of an aketon is that an arrow projecting a few millimeters from the rear face of the *cuir bouilli* may not break the skin of the soldier, and if it does cause a shallow wound the rebound will withdraw the weapon and allow unhindered bleeding to flush contamination.

Figure 14 shows the basic principles of a 'classical' leaf shaped arrow penetrating into hide-based armour. The retarding force comes mainly from the friction as the armour is displaced and grips the sides of the projectile. However, if the armour is to remain intact the compressive forces at the sides of the blade must be balanced by corresponding tensile forces in the material next to the cutting edges. A material such as leather is tough (*i.e.* not prone to cracking), but also flexible so it is easily pushed aside. It is poor at stopping arrows. In contrast to leather, boiled leather is rigid but has very low toughness. It cannot absorb the tensile stresses imposed by a penetrating arrow and cracks run instantaneously from the sharp edges of the blade as the armour fractures under very low stress (figure 15).

The key to successful *cuir bouilli* type armour is to boil rawhide sufficiently to strike a balance between the extreme properties of leather and boiled leather. In practice this means stopping the boiling of rawhide at a point where the rigidity has increased but sufficient toughness remains to prevent brittle fracture. Typically

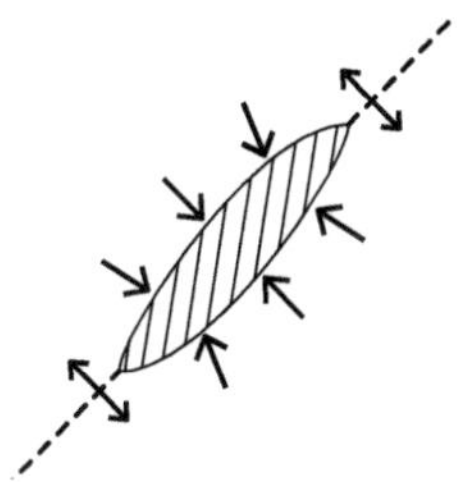

Figure 14. Simplified diagram of the stresses caused by a leaf shaped arrow penetrating into hide-based armour. By Erno Endenburg after Edward Cheshire.

Figure 15. Boiled leather is brittle and shatters when struck by an arrow. Photograph by Edward Cheshire.

this point can occur after some 10 minutes, depending on the thickness of the rawhide (figure 16).

Where hide-based armour has a hard surface layer an additional mechanism comes into play against arrow penetration. The hard coating is very rigid and tightly grips the arrowhead which is trying to penetrate, with the high compressive forces balanced by tension in the hide backing. Under the impact the armour tends to bend concave, further tightening the grip on the arrowhead (figure 17). An arrow strike at 90 degrees, as shown in the figure, would be unusual in battle and the hard coating also tends to deflect missiles hitting at an oblique angle *i.e.* they ricochet.

Figure 16. Tensile stress has caused cracks to start in this rawhide boiled in hide glue, but the fibrous structure of hide has permitted very little opening of the cracks and the arrow has been stopped with 14 mm projection from the rear face. Worn over a padded aketon this armour would have prevented a wound more than a few millimeters deep. Photograph by Edward Cheshire.

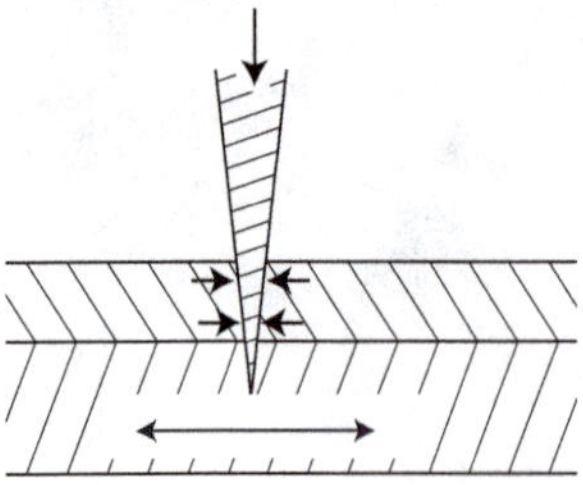

Figure 17. An arrow strike on hard-faced armour. Diagram by Erno Endenburg after Edward Cheshire.

Examples of arrow impacts

Over eighty shots against armours were conducted, mostly employing putative reproductions of *cuir bouilli*, the later examples being hard faced. The following examples have been selected as demonstrating typical features of various types of armour.

Boiled rawhide – trial 45

Armour 05-14 is typical of rawhide boiled in a mixture of hide glue and water, for 20 minutes in this case.[2] The initial thickness was some 4 mm and after boiling the thickness at the impact site is 7.4 mm, although the edges are slightly thicker. The area of the specimen is now less than half that of the original rawhide.[3] The areal density of a piece of armour with a uniform thickness of 7.4 mm would be 8.8 kg/m^2 (the same as iron 1.1 mm thick), while the 'aketon' of linen and cotton added a further 7.8 kg/m^2.

The photographs in figure 18 show the impact of an arrow on boiled rawhide specimen 05-14. In 18a the arrow is in free flight at about 36 m per second and just about to make contact. By 18b the arrowhead has penetrated the rawhide to the maximum extent, but the two are still moving to the right as the aketon is compressed. The arrow and armour are both stationary in 18c. By 18d the arrow and armour have rebounded and the fabric is visible. In this trial the point of the arrow stopped 19 mm behind the rear face of the rawhide. There was no cut into the underlying plasticine, which is used as a substitute for flesh, but a shallow depression about 12 mm in diameter bore the pattern of the fabric. Anyone wearing the armour would have received at least a bruise, with perhaps a small stab-shaped cut if they had been hit on a rigid part of the body over bone. Had the abdomen been struck the peritoneum would probably not have been penetrated. Figure 19 is a Velocity/Time plot of the impact. The arrow is initially travelling at 36 m per second and the armour is at rest. After impact the arrow rapidly slows down as the armour accelerates, until they move together after about 1.6 milliseconds. Eventually their direction of travel reverses as the aketon rebounds (figure 20).

This trial demonstrates the value of a thick aketon in helping to reduce the risk of injury. A further shot at the same armour but with a much thinner aketon caused a 'wound' 19 mm deep in the plasticine. The greater deceleration resulted in a greater projection from the rear of the hide; 24 *cf.* 19 mm; the thinner aketon was insufficient to prevent a wound.

---

2 An initial mixture of 200 grams of gelatine per litre of water was used but the concentration changed as the water evaporated. Gelatin passes completely into solution in water above 49^0 C. At a lower temperature it forms a gel.

3 The specimen measures 11 cm x 8 cm and was hit near the centre. The density is 1.19 gm/ml. The 25 cm square padded 'aketon' comprised 0.55 kg/m^2 of cotton wool and 24 layers of cotton fabric, weighing 0.415 kg/m^2, sandwiched between six layers of linen weighing 0.312 kg/m^2. The plasticine backing was 5 cm thick, packed in an open fronted metal box.

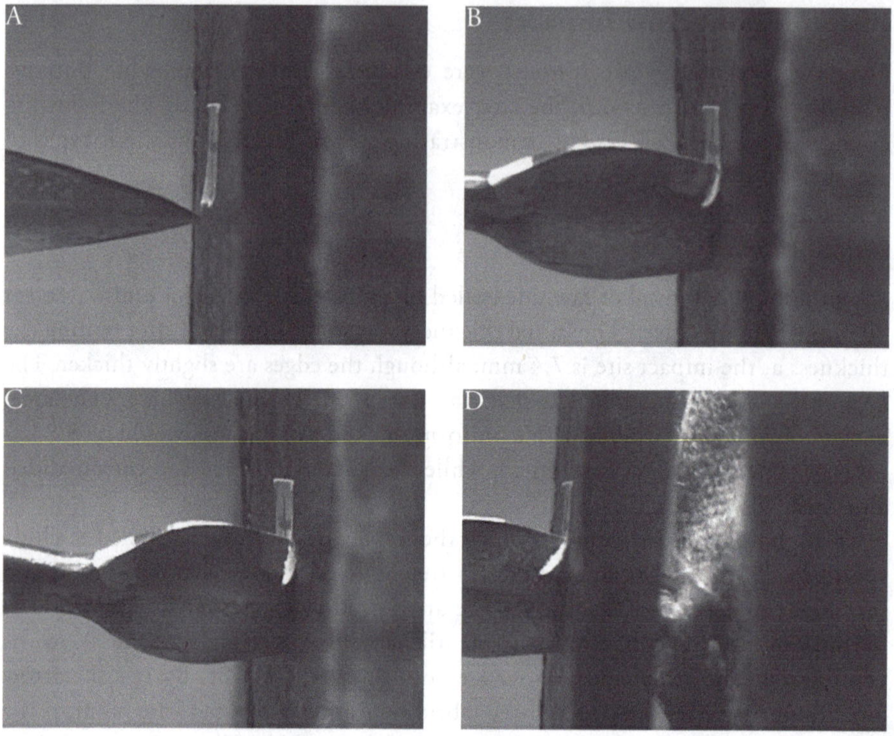

Figure 18. The impact at 36 m/sec and normal incidence of a mild steel headed arrow weighing about 70 grams on replicated rawhide cuir bouilli *over an aketon. The small piece of paper is a photographic focusing aid. Photograph by Edward Cheshire.*

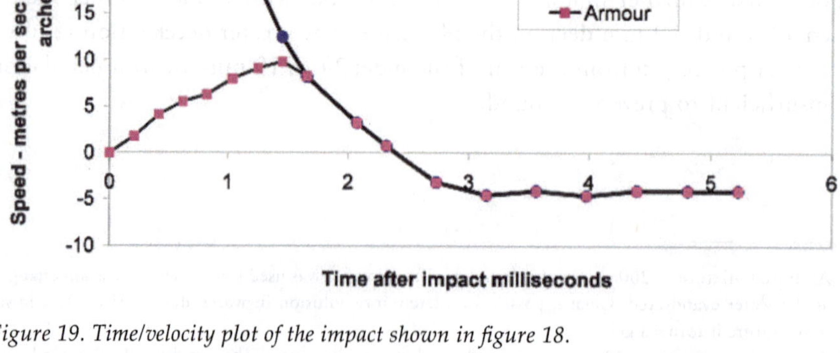

Figure 19. Time/velocity plot of the impact shown in figure 18.

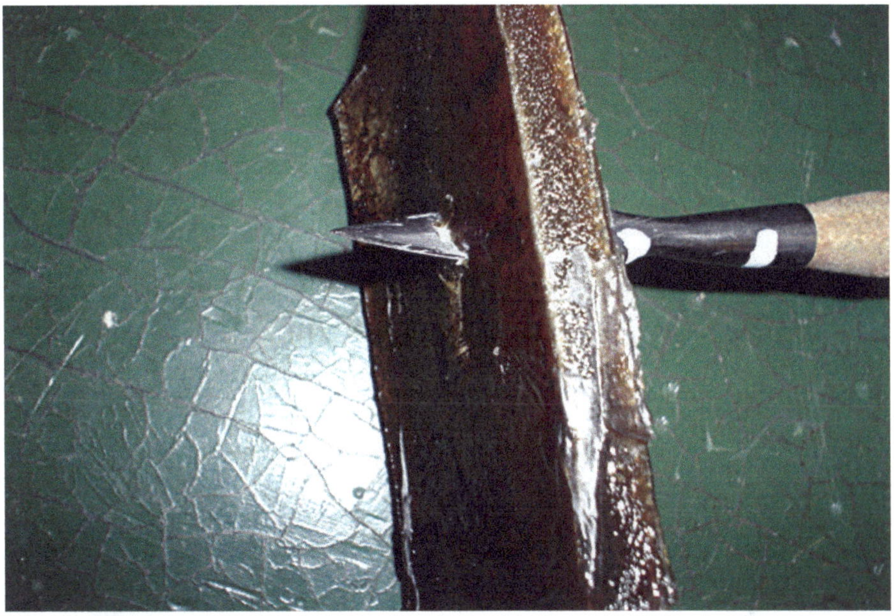

Figure 20. The arrowhead projects 19 mm from the rear face of the armour, but because of the thick padding there was no 'wound' in the plasticine except for a fabric impression 12 mm in diameter. The wearer would have received a bruise and perhaps a wound a few millimeters deep, depending on the impact site. Photograph by Edward Cheshire.

Wax impregnated leather – trial 78

Cuir bouilli is sometimes stated in dictionaries etc. to be wax impregnated leather, but a simple experiment demonstrates this to be improbable. If anything, the wax lubricates the arrow as it penetrates. Armour 06-01 was leather impregnated with beeswax. Being leather rather than rawhide, and therefore not having been heated in water, the thickness and area were substantially unchanged from the original leather. The areal density of the armour and padding was 8.5 kg/m^2 compared with 16.5 kg/m^2 for trial 45 above. Figure 21 shows the arrowhead projecting 88 mm from the rear face of the armour. With statistical correction, to allow for penetration into flesh rather than plasticine, the arrow would have caused a wound over 200 mm deep, if it did not strike bone. The arrowhead would have passed through a limb or be deep in the body and be almost certainly fatal, especially with Medieval medical treatment.

Soft armour - trial 54

Armour 07- 06 comprised three layers of heavy linen fabric (0.52 kg/m^2) and 44 layers of woven silk (0.2 kg/m^2), backed by two layers of cotton wool, to give a total areal density of 11.4 kg/m^2. The leaf-shaped arrowhead penetrated about 80 mm from the impact face to cause a wound in the plasticine 65 mm deep. The corrected wound into flesh would probably have been well over 100 mm deep (figure 22).

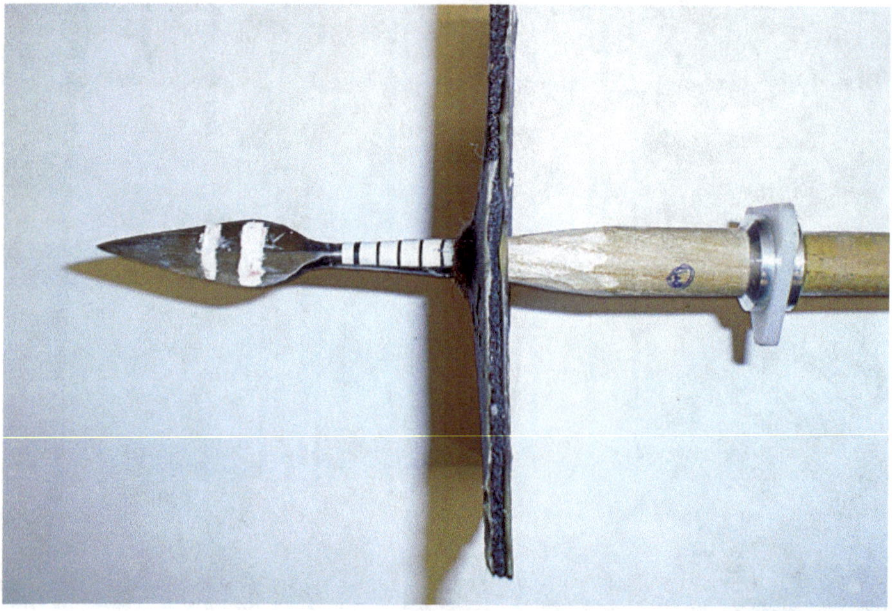

Figure 21. The arrowhead projects 88 mm from the rear face of leather impregnated with beeswax and would have penetrated much further into flesh rather than plasticine (about 200 mm if it did not hit bone). Photograph by Edward Cheshire.

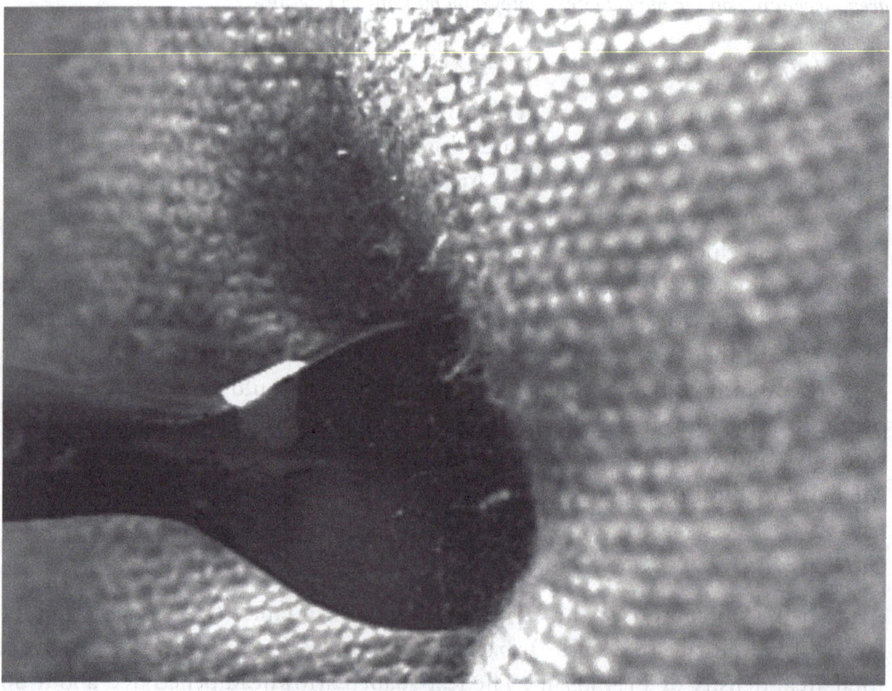

Figure 22. The arrowhead penetrating soft armour 07-06, 1.4 milliseconds after initial contact. The fabric is coming under tension and the fibres must be both absorbing strain energy and sliding across each other to admit the blade, which is nearly twice the width of the final slit in the top layer. Photograph by Edward Cheshire.

This trial demonstrated two typical phenomena observed with some fabric armour. Firstly, the arrowhead was 19.8 mm wide but had created a permanent slit only 10 mm wide as it passed through. This is probably because the fibres stretched under the tension of the impact and also slid across each other to admit the blade. Secondly, the silk, which has exceptionally good strain energy absorption properties, gripped the waist of the arrowhead so tightly that it was difficult to extract it. Provided that an arrowhead did not penetrate so far as to be lost in the body, the grip of the silk would be an aid to the surgeon during its extraction.

Hard faced boiled rawhide - trial 58 (set at an angle of 30 degrees to the line of flight)

The arrowhead ricocheted off the hard surface (crushed marble in hide glue) leaving only a minor scratch 18 mm long, but the arrowhead was permanently bent sideways by the impact. Anybody wearing this armour would be unharmed (figures 23 & 24).

Ram's horn - trial 14

This armour specimen comprised one layer of heavy linen fabric (0.52 kg/m^2) over a piece of flattened ram's horn, behind which were further layers of linen with cotton wool. The horn weighed 15 kg/m^2 and the total areal density was 20.82 kg/m^2. The armour approximated to a panel of horn worn in a brigandine. The

Figure 23. The arrowhead approaching from the left is bent downwards by the impact, 0.29 milliseconds after the initial contact. A small amount of marble facing is peeling off just below the point of the arrowhead. Photograph by Edward Cheshire.

Figure 24. The marble facing received a minor scratch 18 mm long (visible running horizontally above the centre of the picture). Photograph by Edward Cheshire.

arrow was stopped effectively, penetrating some 10 mm into the 11.7 mm thick horn. The square section arrowhead was chosen because it had greater resistance to buckling than a leaf-shape, but it was nevertheless permanently bent by the impact. Anyone wearing this armour would have received no penetrating wound (figure 25).

Some medical matters

A feature of hide-based armour which became apparent as the trials progressed is the way in which a projectile such as an arrow is stopped. There is often sufficient penetration to cause a shallow wound, although injury can be limited by balancing the properties of the armour and aketon with the physiology of the underlying body. A buttock can tolerate some deflection and even penetration without significantly impairing the ability of the warrior to fight on, but the same cannot be said of the knee or elbow. Some of the earliest plate armour was designed to protect exactly these areas, where mail was found wanting as mail needs to deflect before it can go taut and protect.

The surgeons who left the most useful records of arrow wounds were employed by the Army of the United States in the campaigns against the indigenous population during the 19th century AD. These men were required to report on casualties and had good anatomical skills. They mostly practised shortly before the antiseptic discoveries of Lister were employed but the arrows they encountered were similar to those on a Medieval battlefield, with similar ability to wound and

Figure 25. Ram's horn offers good resistance to penetration by arrows. Photograph by Edward Cheshire.

kill. These surgeons recorded sufficient casualties to give a clear picture of the danger of arrow wounds, in contrast to the few anecdotal comments from early times. Modern records from countries such as India, New Guinea and Nigeria are of limited value because modern advances such as antibiotics and X-rays must distort survival rates compared with the Medieval experience.

If it does not kill quickly, the greatest danger of an arrow wound comes from losing the arrowhead within the body or deeply embedded in bone. The Hollywood image of a wounded hero pulling out an arrow complete with the head is cinematic licence. In reality the head was usually left behind to fester and kill. In none of the trial shots at *cuir bouilli*, hard faced or not, did the arrowhead pass completely through the armour. It either ricocheted or was trapped in the armour, such that it would have been extracted when the armour was removed. This is an important feature, which was not shared by boiled or wax impregnated leather.

Bill (1862, 366-367) relates the danger of arrow wounds thus:

> *"Let us suppose a case to illustrate and explain our meaning. An arrow is shot at a man at a distance of fifty yards. It penetrates his abdomen, and without wounding an intestine or a great vessel, lodges in the body of one of the vertebra. The arrow is grasped by the shaft by some officious friend, and after a little tugging is pulled out. We said the arrow is pulled out. This was a mistake; it is the shaft only of the arrow that is pulled out. The angular and jagged head has been left buried in the bone to kill – for it surely will – the victim [...] Experience has abundantly*

> *shown, and none know the fact better than the Indians themselves, that any arrow wound of chest or abdomen, in which the arrow-head is detached from the shaft and lodged, is mortal. From this we conclude that the danger* peculiar *to all arrow wounds is,* that the shaft becoming detached from the head of an implanted arrow, leaves this so firmly imbedded in a bone that it cannot be withdrawn, and that, remaining, it kills. *It is not possible with forceps to extract an arrow-head so lodged (if lodged deeply) throwing aside the difficulty of discovering and the danger of searching for it. The blades of forceps long enough for this purpose (supposing the foreign body deeply lodged in the chest) would bend too readily with the force required for the removal of the missile. The greatest force is sometimes required for the extraction of an arrow-head so lodged. We have seen an arrow shot at a distance of one hundred yards, so deeply imbedded in an oak plank, that it required great force, applied by strong tooth-forceps, to remove it [...]" (Emphasis [regular font] in original).*

Warming to his task, Bill (1862, 373-374) continues, with particular reference to the limbs:

> *"Keeping in view the real cause of dread in arrow wounds – to wit, the lodging and fixation of the iron head in a bone [...] the arrowhead may be detected, and if detected, extracted with facility. In any event, frequent search should be made for it. Generally we will at length succeed in removing the foreign body; but all things else failing, as time, strength and patience may, we must operate. The incisions should be large and free – boldness rather than prudence governing our actions. We might as well cut the patient's limb up until we do find the arrow-head, for if it is left, amputation will be necessary, and worse than this can hardly ensue from the 'cutting up' we have advised. We would, if we undertook such an operation, make up our mind to find the arrow-head, even if it were necessary to tear up every fasciculus of every muscle of the injured member. [...] removal should always be effected as soon as possible after the receipt of the injury,* and the greatest care taken in doing so not to detach the shaft from the head of the arrow." *Emphasis [regular font] in original).*

If it does not kill quickly by injuring an organ or by loss of blood, an arrowhead left in the body will most probably cause fatal infections such as tetanus and gas gangrene, especially if the fight is over cultivated ground harbouring anaerobic bacteria.

Penetration results and comparisons

Not face-hardened

Figure 26 summarises the results of shooting trials against various types of non-face-hardened armour. The wound depths shown have been modified from the penetrations into plasticine by a factor to give the expected penetration into flesh. Trials by others involved shooting various arrows into recently deceased pigs to observe the depths of wound and the author calculated the appropriate factor

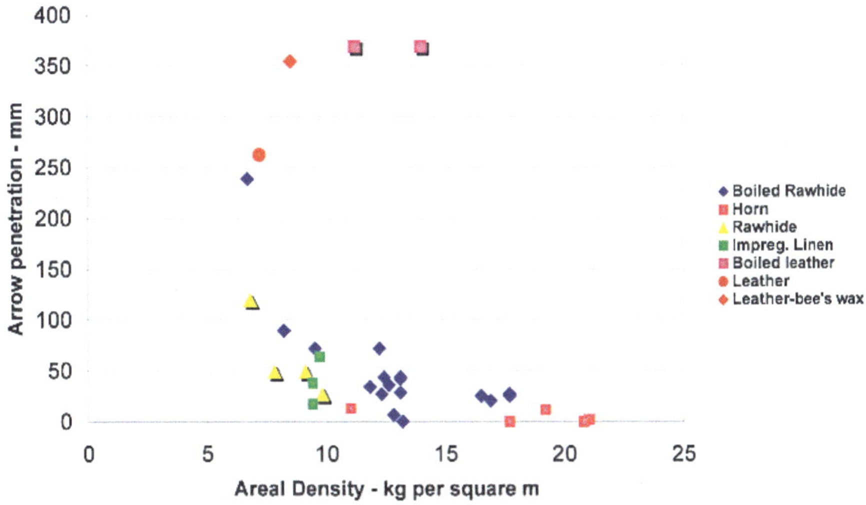

Figure 26. Comparison between depths of 'wound' resulting from arrow strikes on various types of armour, plotted against Areal Density (with wound depth correction).

(between one and four) to apply (Karger 1998, 495-501). Where the arrowhead barely penetrated the armour little or no correction was necessary, unlike 'wounds' several centimetres deep into plasticine which demand substantial correction.

Those materials generally suggested in the literature as being '*cuir bouilli*', *i.e.* boiled leather and wax impregnated leather, offer such poor protection against arrows that it is highly unlikely that they were employed in any battle other than in modern re-enactments. The best material is ram's horn, which would be excellent for small panels such as in a brigandine. Cow horn was not tested but good results should be expected from this also. Rawhide, and especially boiled rawhide, offers reasonable resistance to penetration and a soldier wearing armour of this type would stand a good chance of suffering minor wounds or even none (the penetrations shown are measured from the rear face of the hide, which would normally be worn over a padded aketon).

As might be expected the heavier armour offered generally better protection. The impregnated linen performed well at a relatively light weight and it would be straightforward to increase the areal density, and hence the protection, by simply adding more layers of linen. Hide based armour is based on natural organic material and a degree of scatter is to be expected in the results. For example, one piece of boiled rawhide (the lightest) performed relatively poorly. Figure 27 is a plot showing only boiled rawhide, both un-faced and hard-faced (the vertical scale of figure 27 is four times that of figure 26). The un-faced material exhibits a reasonable fit to a curve such as might be expected, again with heavier armour giving the better protection.

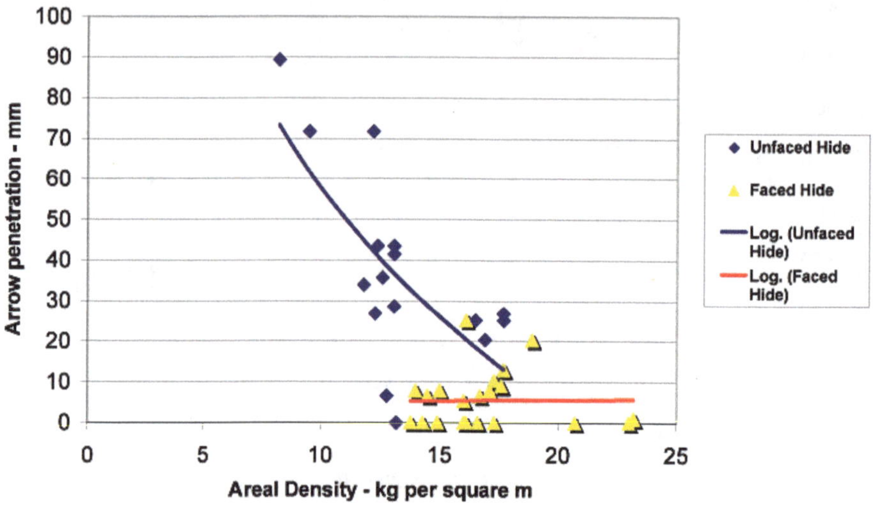

Figure 27. Graph of Penetration against Areal Density. Unfaced boiled rawhide and faced boiled rawhide.

Face-hardened

Figure 27 shows the good resistance of hard-faced boiled rawhide to penetration by arrows. The mechanism is probably as shown in figure 17 (and figure 14) but there is no obvious correlation with the areal density. The log plot gives an approximate horizontal line. It may be that the hard face does most of the work gripping and stopping the weapon as long as the backing has sufficient tensile strength and toughness to keep the armour substantially intact. If this condition is met, adding further material to the backing would have little effect other than increasing the mass. A reasonable adhesive strength is also necessary if the hard face is not to spall and fail to grip the arrowhead.

Merit Indices

Merit Indices give some indication of comparative efficacy (table 3). Here the ratio Depth of Wound/Areal Density is used, normalized to give boiled rawhide the numerical value 'one' (a low numerical value is 'good'). This ratio has been chosen as a comparison of how much protection a soldier might expect from various materials for a particular weight of armour to be carried into combat.

To resist arrow penetration the best of the materials tested was found to be ram's horn. This has to be cut, heated and flattened to obtain small panels such as might be used in a brigandine. It is not available in large pieces. Hard faced boiled rawhide had a merit index of 0.3, compared with the 1.0 of un-faced boiled rawhide. Although considerable skill would be necessary, it would be possible to produce a large piece of armour such as a cuirass from these materials, as implied by the coriaceous name. Figure 28 shows a stone effigy in the Abbey at Pershore in Worcestershire which is believed to be wearing just such armour.

Because of the small number of samples tested, only approximate values have been allocated to leather, waxed leather and boiled leather: conservatively they are all quoted as being more than six. Thus a soldier wearing armour of any of these materials might expect to receive an arrow wound say 15 to 20 times deeper than a comrade clad in hard faced boiled rawhide, without considering his aketon or ricochet or impact with bone; *i.e.* 6 / 0.3 = 20.

Material	Merit Index
Horn	0.2
Hard faced boiled rawhide	0.3
Boiled rawhide (datum)	**1.0**
Impregnated linen	1.5
Rawhide	2.2
Leather	>6
Waxed leather	>6
Boiled leather	>6

Table 3. Normalised Merit Indices of Non-metallic Armours (calculated from the ratio: Depth of Wound/Areal Density).

Figure 28. Stone effigy of a knight in Pershore Abbey, Worcestershire. The right arm of the prone knight is at the bottom of the picture and his chest, covered by a fabric surcoat, is at the top. The front and back halves of the cuirass are buckled together by leather straps through slots in the cuir bouilli. Photograph by Edward Cheshire.

Conclusions

The material known as *cuir bouilli* was widely used as armour during the Middle Ages but there has been no consensus about how the material was made. Researchers have tended to assume '*cuir*' to be only 'leather' (which would be correct in modern French) and '*bouilli*' to mean 'boiled'. Simple experiments show that boiling leather in water produces a hard but extremely brittle material which is prone to shatter and effectively useless as armour. In response to this discovery, workers have assumed that perhaps *bouilli* might only mean treatment in hot, but not boiling, water, but this approach has not been successful either. Old French '*cuir*', in a variety of spellings, meant not only leather but also rawhide, skin and even flesh. Experiments to boil rawhide demonstrated that this technique does indeed produce a material which stops arrows effectively. Armourers would require considerable manual skill to accurately manufacture large items such as a cuirass but the basic technology is straightforward. *Cuir bouilli* made by boiling rawhide is functional armour and the principle conflicts with neither the interpretation of the Medieval term or the scant literature.

The theory and practice of making boiled rawhide armour has been described in some detail, with particular attention to the effect that the time of boiling has on the material. The primary mechanism of arrow penetration into boiled rawhide (figure 14) shows how the material needs a high rigidity, to avoid being pushed aside, coupled with sufficient tensile strength and fracture toughness to prevent significant cracking. Of the hide-based materials only rawhide boiled for the optimum time, about ten minutes depending on the thickness, approaches the required balance of properties. Leather, whether only tanned or impregnated with wax, does not grip the weapon sufficiently tightly, whereas boiled leather is too brittle and prone to shatter. Natural rawhide has reasonable resistance to penetration, which is improved by adequate boiling but then deteriorates if the boiling is continued for too long.

A hard surface layer, typically of crushed rock in hide glue, further increases resistance to penetration (figure 17). The coating grips the weapon very tightly and retards it, while the balancing tensile force is provided by the backing. In practice, if the arrowhead does not ricochet, the mechanisms shown in figures 14 and 17 occur simultaneously in a strike on hide-based armour with a hard surface.

The mechanical tests gave useful information but also threw up some complications. The tendency of rawhide to progressively 'case harden' from the outside surfaces when it is boiled causes particular problems, because the properties of the skin are different from the core and the relative proportions vary with time. Nevertheless the tables of relative tensile strength and Charpy impact testing show how boiling changes the material properties. As the tensile strength reduces, so do the impact resistance and fracture toughness. The increase in stiffness, needed to grip a penetrating weapon, goes hand in hand with the increase in hardness and is also indicated indirectly by the impact resistance. The data confirm how the boiling of rawhide modifies the properties to meet the figure 14 parameters.

Armour is always a compromise between protection, weight and cost, but the prime purpose of battle armour is to reduce the likelihood of injury, and especially of a crippling wound which prevents the warrior from fighting on and defending himself ('parade' armour may be designed mainly to impress). The mechanism by which hide-based material stops arrows usually involves some penetration into the armour and perhaps a minor wound, but in no trial shot with boiled rawhide did the arrow pass right through. The arrowhead was tightly gripped by the armour and would have been removed with it, which would have had immense medical implications on a medieval battlefield. The outcomes of arrow wounds have been discussed and, taken in conjunction with the shooting trials, serve to emphasise the value of boiled rawhide compared with leather, whether natural, boiled or wax filled. If a medieval soldier managed to live through the combat, but with an arrowhead below the surface of his body he presented the surgeon with a dilemma. If the shaft of the arrow was projecting through the armour and still accessible the surgeon would have to decide whether or not to pull it out or cut it off, because he could probably not remove the armour with the shaft *in situ*.

The photographs that are shown above with arrows embedded in *cuir bouilli* are consistent with accounts from the crusades of arrows penetrating Frankish armour but frequently not causing injury. The image of the porcupine was sometimes used to describe men or animals who had been under Turkish attack (Small 1997, 81) although arrows sticking in mail could appear similar (Joinville 1983, 261). About eighty trials involved shooting arrows at various types of armour, mostly hide-based, and some with a hard face but including ram's horn, impregnated linen and soft armour of cotton, silk and linen. Most armours were mounted in front of a simulated padded aketon which absorbed energy and significantly influenced the depth of any wound into the plasticine backing. Representative shots have been described, together with discussion of the observations. The outcomes are summarised in the form of graphs of Wound Depth against Areal Density (figures 26 & 27). The complete range of materials shot at is shown in figure 26. An arrow hit on boiled leather would shatter the armour and penetrate perhaps 350 mm into flesh, and armour of natural leather or wax filled leather would offer only slightly more protection, with penetration into flesh exceeding 250 mm for the areal densities shown. Rawhide gives improved results and horn is supreme. Impregnated linen gives good protection and the thickness, and therefore the areal density, would readily be manufactured in various thicknesses by adding more or fewer layers. Of the hide-based armours, boiled rawhide offers a good combination of resistance to penetration with a given areal density. The thickness and areal density are both increased by boiling. Figure 27 shows the dramatic effect of a hard surface on boiled rawhide (the boiled rawhide data are the same as the lower values in figure 26 but the vertical scale is different). In most cases the arrow failed to penetrate more than a few millimetres and would not have caused a wound through an aketon. For reasons that are not fully understood, the penetration/areal mass relationship for a hard surface armour is not straightforward like that of un-faced boiled rawhide, the log graph of which is a curve as would be expected.

The penetration performance of the various armours is complex and influenced by a wide range of parameters, such as arrow velocity and weight, arrowhead geometry, armour weight, angle of arrow to armour at impact, hide thickness and method of armour manufacture. Inconsistency in these is responsible for some of the scatter of points on the graphs and could undermine the validity of the outcome of the research if the results were not so clear cut. To simplify the comparison between the hide based armours in particular, the Merit Index principle has been employed, based on the ratio Penetration/Areal Density (this ratio is itself of value only if the areal densities are roughly similar). The values are normalised, with boiled rawhide allocated a datum value of one (table 4).

A soldier hit by an arrow while wearing armour of unmodified vegetable tanned leather, wax filled leather or, especially, boiled leather could expect to receive a wound at least six times deeper than if he were clad in boiled rawhide of similar mass. The addition to boiled rawhide of a surface of crushed rock in a hide glue matrix would reduce the wound depth by a further factor of about 3 *i.e.* by a total factor of 6/0.3 = say 20.

Cuir bouilli armour was widely used during the Medieval Period. Despite the claims by others that it was leather, wax filled leather or boiled leather, this seems unlikely as it is clear from the discussed experiments that these were not up to the job. Rawhide offers reasonable protection but it is not compatible with the little that is known about *cuir bouilli*. It is, therefore, suggested that *cuir bouilli* was in fact boiled rawhide, and that, in Palestine at least, it was much improved by the addition of a surface layer of material such as crushed rock, probably in a matrix of hide or fish glue or similar liquid with setting properties.

References

Newey, C. & Weaver, G. (1990). *Materials principles and practice.* Milton Keynes: Open University.

Anfield, G. (1982). *Traditional crafts in Britain.* London: Reader's Digest Association.

Bergman, C.A., McEwen E. & Miller, R. (1988). Experimental archery: projectile velocities and comparison of bow performances. *Antiquity*, **62**. 658-670.

Bill, J.H. (1862). Notes on arrow wounds. *Medical Record* (U.S.A., Oct.).

Hide-based Material	Merit Index
Hard faced boiled rawhide	0.3
Boiled rawhide (datum)	**1.0**
Rawhide	2.2
Leather	>6
Waxed leather	<6
Boiled leather	<6

Table 4. Normalised Merit Indices of Hide-based Armours.

Blair, C. (1979). *European armour – circa 1066 to circa 1700.* London: Batsford.

Cameron, E.A. (2000). *Sheaths and scabbards in England AD 400-1100.* Oxford: Archaeopress.

Catlin, G. (1973) [1841]. *Letters and notes on the manners, customs and conditions of the North American indians.* New York: Dover Publications (Letter No 30. Mouth of Teton River, Upper Missouri, 241).

Chard, T. (1940). Did the first Spanish horses landed in Florida and Carolina leave progeny? *American Anthropologist, New Series* **42**, 1: 94.

Cheshire, E.J. (2010). *Non-metallic armour prior to the First World War.* Unpublished PhD. Thesis: University of Reading.

Courtney-Green, P.R. (1991). *Ammunition for the land battle.* London: Brasseys.

Dobson, C. (2003). As tough as old boots? A study of hardened leather armour. Part 1: Techniques of manufacture. In: Dobson, C., (eds.) *Art and arms, Florence, city of the Medici. Conference proceedings, May 30th to June 1st 2003, Florence.* Clare: Dobson. 79-99.

Everson, T. (2004). *Warfare in ancient Greece.* Stroud: Sutton.

Foulkes, C. (1912). *The armourer and his craft.* London: Methuen.

Froissart, Sir J. (1874). *Chronicles of England, France, Spain and the adjoining countries.* Trans. Johnes, T., Book 4, Vol 2, 471. London: George Routledge and Sons.

Gustavson, K.H. (1960). The behaviour of teleostean and bovine collagen towards specific agents indicating the presence of covalent crosslinkages in bovine collagen. In: Ramanathan, N., (ed.). *Collagen. Proceedings of a symposium sponsored by the Central Leather Research Institute, November 29-30, Madras.* New York: John Wiley. 279-304.

Hindley, A. (2000). *Old French-English dictionary.* Cambridge: Cambridge University Press.

Joinville, J. de (1983). *The life of Saint Louis. Chronicles of the crusade.* Harmondsworth: Penguin.

Josephus, (1959). *The Jewish War.* Trans. Williamson, G.A. Harmondsworth: Penguin.

Kelham, R. (1978) [1779]. *A dictionary of Norman or old French language.* Wakefield: Tabard Press.

Kinloch, A. J. (1987). *Adhesion and adhesives.* London: Chapman and Hall.

Kōzan, S. (1962)[1800]. *The manufacture of armour and helmets in sixteenth century Japan.* Revised and edited by Robinson, H.R. London: Holland Press. 59-60.

LaRocca, D.J. (2000). An approach to the study of arms and armour from Tibet. In: *Royal Armouries Yearbook,* Vol. **4**, 1999. 113-132.

Lingwood, R. (1990). Heat assisted forming. Designer leatherworker. *Journal of the Association of Designer-Leatherworkers,* **2**. 8-12.

Nicolle, D. (2008). A Medieval Islamic arms cache from Syria. *The Journal of the Arms and Armour Society* **XIX**, (4). 153-163.

Nicolle, D. (2002). *A companion to Medieval arms and armour.* Woodbridge: Boydell Press.

Ramanathan, N. & Nayudamma, Y. (1960). The effect of hydrothermal shrinkage and recovery on the optical birefringence of tanned collagen fibers. In: Ramanathan, N., (ed.). *Collagen. Proceedings of a symposium sponsored by the Central Leather Research Institute, November 29-30, Madras.* New York: John Wiley. 453-459.

Richardson, T. (1999). Ballistic testing of historical weapons. *Royal Armouries Yearbook,* Vol. **3,** 1998.

Robinson, H.R. (2002). *Oriental armour.* New York: Dover.

Simpson, J.A. & Weiner, E.S.C. (1989). *Oxford English Dictionary,* 2nd. ed. Oxford: Clarendon Press.

Singer, C. (1957). *A history of technology. Vol. II.* Oxford: Clarendon Press.

Small, R.C. (1987). *Crusading warfare 1097-1193.* Cambridge: Cambridge University Press.

Tulloch, S. Ed. (1993). *The Reader's Digest complete wordfinder.* London, Readers Digest Association.

Waterer, J.W. (1981). *Leather and the warrior.* Northampton: Museum of Leathercraft.

Weidman, G., Lewis, P. & Reid, N. (1990). *Structural materials.* Milton Keynes: Open University.

Witnauer, L.P. (1960). The relationship between shrinkage of hide and the crystal- liquid transition of collagen. In: Ramanathan, N., (ed.). *Collagen-Proceedings of a Symposium sponsored by the Central Leather Research Institute, November 29-30 1960.* New York: John Wiley. 441-451.

Bespoke vellum: some unusual requests

Laura Youngson Coll

Art and craft: an introduction to making

In this contribution the question 'Why leather?' is addressed through reflecting on the author's training and practice as an artist and skilled leatherworker. The context for this discourse is the use of leather and vellum as a decorative material in contemporary interiors. An account of three such interior projects will examine a number of issues exploring the question 'Why leather?' including the production of specific leathers, the material qualities of leather, the knowledge and practice of the craftsperson, the dialogue between designer and craftsperson, and the demands of the interiors industry. The author's decorative art work in leather will be examined, exploring the ramifications of working as an independent artist without the parameters of a set design brief. This, in turn, will demonstrate the approaches adopted to fully explore the material and aesthetic qualities of leather, further contextualizing the question 'Why leather?' In addition to the author's experience, two craftspeople have been interviewed whose practice is predominantly in interior leatherwork, investigating the diverse backgrounds which feed into this highly specialized area of craft. Ada Nuottajarvi spent five years training on a Master of Arts degree in Special Techniques (*Erityistekniikat*) at Kuopio Academy of Design in Finland and Caroline Grappe took a degree in Book Art and Bookbinding at London College of Printing (now London College of Communication) in London.

The projects

The three projects that will be discussed are the creation of a vellum corridor (Project One), the making of a shagreen table (Project Two) and goatskin vegetable-tanned leather (Project Three). The spectrum of work produced within the bespoke interior leather industry is extensive (figure 1). Innovation, exclusivity, material (figure 2) and labour time are all equated to produce viable work: all under the auspice of luxury bespoke interior leatherwork. Vellum and shagreen, the predominant materials that will be discussed, are not by strict definition leather as they are cured rather than tanned. Within the interior design industry they are considered specialist leathers, allowing a looser terminology. For the purpose of

Figure 1. From left to right: 'Russia Hide' door by Gavin Rookledge ('Russia' Hide is more usually termed 'Russia leather': in 1786 The Metta Catharina, a Russian ship heading to the Mediterranean from St. Petersburg became wrecked on the Plymouth Sounds. Its cargo included reindeer hides tanned with willow bark and birch oil and embossed with handmade cross hatched pattern. In 1973 divers found many bundles of hides scattered on the sea bed. Due to prolonged immersion in black mud they are remarkably well preserved, Skelton 2010, 235-257). Courtesy/copyright of Rook's Books Ltd. Photograph by Laura Youngson Coll; Handrail in upholstery leather with machine and hand stitch detailing. Courtesy/copyright of Harcourt Leather Design Ltd. Photograph by Harcourt Leather Design Ltd; Making a vellum cabinet with decorative pared joins. Courtesy/copyright of Rook's Books Ltd. Photograph by Laura Youngson Coll; Large wall in chrome tanned silver upholstery leather with piping detailing. Courtesy/copyright of Harcourt Leather Design Ltd. Photograph by Harcourt Leather Design Ltd.; Decorative wall panels with leather covered laser cut pattern. Courtesy/ copyright of Harcourt Leather Design Ltd. Photograph by Harcourt Leather Design Ltd.

Figure 2. Some unusual materials used for interiors. Left to right: Faux shagreen (shagreen pattern embossed on upholstery leather); faux crocodile (as previous); woven horse hair; embossed vegetable tanned cowhide; eel skin (sewn together); sturgeon skin. Photographs by Laura Youngson Coll.

this paper these materials are included under the definition of leather. To put these projects in context, the environments and working ethos in which they were made will be discussed.

The projects specified are the work of two different companies. Project One is the work of a company which originated as an artisan bookbinder and then moved into interiors. The work is all handmade, no machine work is used, and many of the techniques are appropriated from traditional bookbinding skills. The client, usually the designer, establishes the exclusivity of the work. In turn, the designer's clients are highly confidential.

Conversely, the company which completed Project Two works for a much broader range of designers and architects on a greater diversity of projects. This gives a wider perspective on the industry. The range of work produced by the company encompasses large upholstered wall panels to intricate laser cut and hand stitched detailing. Similarly, the range of projects varies from new-build mansions to historically significant restoration.

Each company chooses to use different types of leather and working methods dictated by image, attitude, skills, technological influence, economics and ethics. All these factors have a fundamental bearing on the ways in which the makers, within the company, work with leather. This will be further examined in the next section, expanding the question 'Why leather?'

Project One: the vellum corridor

Handmade processes and the appropriation of traditional techniques

Mapping out the project

Project One involved a significant number of vellum ribs to be installed in a corridor (figure 3). A pattern of decorative joins were mapped out across a scale drawing of each wall, in which these ribs were the component parts. Once this templating system had been established, the pieces could then be worked on individually, as each one would be referenced as a component in the complete wall. Most of the initial process of this project was logistical, with a great deal of laborious checking and marking up, but the preparation and application of the vellum required was more technically specialized.

Figure 3. Components of the vellum rib corridor. From left to right: Wall ribs; plug socket and handle details; door ribs. Courtesy/copyright of Rook's Books Ltd. Photographs by Laura Youngson Coll.

The choice of vellum in this project is ultimately the designer's decision based on the client's expectation. However, it is the maker who initiates the process and suggests the possibility of using vellum to the designer. Here, different aspects of the making process will be discussed in relation to this project and vellum, creating a discourse around the question 'Why leather?' The focus will be on the perspective of the maker but the decisions and influence of the designer, the clients and the wider interiors industry are being looked at as well. Understanding the methods used to cure the skin and produce vellum is the first stage of this exploration.

The material: vellum

Vellum is predominantly goat or calf skin but can be prepared from a variety of skins. It is not tanned like leather but cured: cleaned, stretched on a frame and scraped with a hemispherical knife. This wet and dry process creates a tension in the structure rendering the finished skin with a slight rigidity. The resultant skin is fine and smooth on the upper surface. It usually is bleached, resulting in cream or whitish appearance with hair follicles, vein patterns and scars sometimes visible. Alternatively the vellum is dyed or left as unbleached (known as supernatural by William Cowley Ltd.) creating patches of deepened colour and emphasising hair follicles (figure 4).

The producers William Cowley Ltd. are the only company left in the United Kingdom producing vellum through traditional methods. Vellum, or parchment, has traditionally been used for manuscripts and is still used as a material on which to record Acts of Parliament. Botanical illustration and painting also have a long tradition of using vellum as a canvas (Rix & Sherwood 2008, 30-31). This is seen in many of Nicholas Hilliard's Tudor miniatures: the translucent quality of the skin offers the paint a unique illuminating quality (Victoria and Albert Museum P.26-1975). Carlo Bugatti, an architect and designer from the late 19th-mid 20th century, championed the use of vellum in his theatrical furniture (Lucie-Smith 1979, 160).

Working with vellum: techniques and problems

Given its material longevity and relative strength, the contemporary world of interior makers and designers is often wary of vellum. There are a number of problems associated with its application to panels and furniture. If the application is not made correctly air bubbles can become trapped under the skin. The stretching process involved in its initial preparation, and the wet gluing process in which it is applied to a surface, means that the skin retracts as it dries. Due care has to be taken when making joins or making cuts to shape corners: insufficient material can lead to gaps (figure 5). The strength of the contraction during drying can also warp thin wood. The other concern associated with gluing vellum to a surface is the translucency of the skin. With white and cream vellum the prepared surface has to be pristine white, every little hair and speck of dust has to be removed. This needs to be done before the gluing process begins, otherwise this matter is visible under the skin once it is glued down. Two of the main working processes involved in the vellum corridor are outlined below.

Figure 4. Different vellum. From left to right: white vellum; hazelnut dyed vellum; biscuit dyed vellum; 'supernatural' unbleached vellum. Photograph by Laura Youngson Coll.

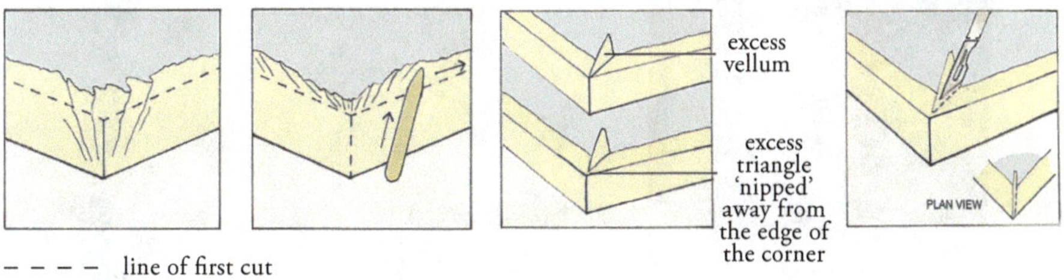

Figure 5. Extract from a training manual for vellum application. Author Laura Youngson Coll. Copyright of Rook's Books Ltd.

Starch paste gluing

The process used to apply the vellum to a primed surface is a wet gluing process. A polyvinyl acetate (PVA) and starch paste is worked into the underside of the vellum with a large paste brush. This is repeated periodically over a period of around twenty minutes until the skin becomes flexible. It is then positioned on the panel and smoothed over with a bone folder, removing most of the glue from under the skin in the process, and resulting in a smooth hard surface. This is a process derived from bookbinding techniques in which vellum is used as an alternative to leather, and applied to the boards of the book.

Paring

Paring is a process of thinning down the edge of leather or vellum with a surgical scalpel or paring knife (figure 6). This is done on a blade resistant surface such as glass or marble. Different degrees of thickness can be achieved in order to make a smooth join when one piece is placed upon another. In much of the decorative and furniture work this join is made extremely fine, eliminating any blunt edge. The overlap made by adjoining pieces causes a darker line; this can be used as a decorative device making anything from angular patterns to smooth curved lines (figure 7).

Why vellum? The maker's choice

For the maker all the technical factors outlined contribute to a choice to work with vellum. Aesthetically, it can produce anything from fine, subtle white surfaces to complex patterned surfaces formed of hair follicles and scars. Vellum can be manipulated, using the wet gluing process, and moulded around surfaces, taking on the shape of the object it is applied to. It has potential to become translucent if it is sufficiently thinned or if light is directed through the skin. The time

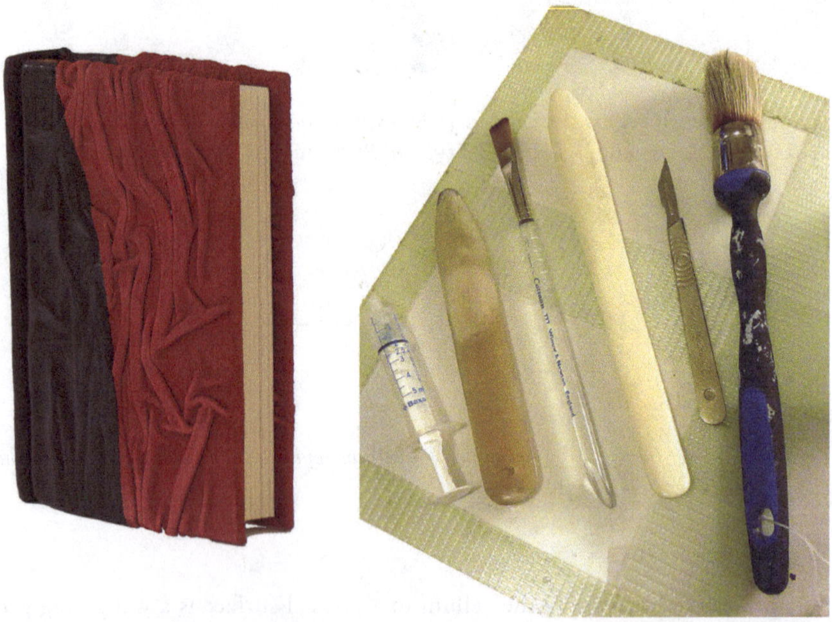

Figure 6. Left: Book with new leather binding. Right: Tools used in bookbinding and fine interior work. From left to right: Polyvinyl acetate (PVA) syringe; horn bone folder; small brush; bone folder; scalpel; paste brush; cotton wool; all resting on paring glass. Photographs by Laura Youngson Coll.

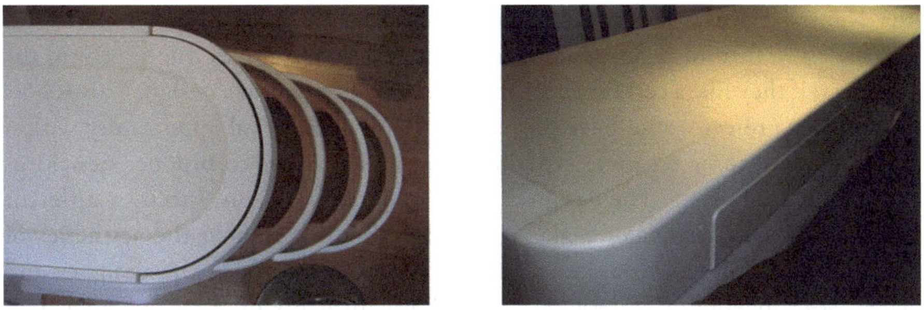

Figure 7. White vellum furniture showing decorative pared joins. Courtesy/copyright of Rook's Books Ltd. Photographs by Laura Youngson Coll.

consuming and specialist skill involved provides technical and aesthetic challenges, and is therefore a measure of creativity of the maker. The question 'Why leather?' becomes one of duality; the maker chooses vellum for a different reason to the designer but ultimately both must meet the clients' expectation. In the next section we will explore the consequential relationships between these parties and how this affects a maker's approach to the use of vellum. Firstly, however, we will take an initial look at the designer's perspective.

Why vellum? The designer's choice

The designer's choice of vellum on this project resulted from a perception of the material as unusual and aesthetically successful in a particular decorative scheme. The maker in this case devised a successful way to apply and mould the vellum onto wooden panels. The success of this technique creates exclusivity where others found the material difficult to work with. The imaginative use of pared joins further enhanced this technique aesthetically and practically. All these factors, the current scarcity of quality vellum production, the prestigious legacy of vellum, its use in things such as Acts of Parliament, the high cost of the material and the speciality of application methods make the material appealing to a designer whose clients wants exclusivity. The appeal of using a skilled maker, who has an in-depth knowledge of their material and has a greater contribution than just technician, becomes apparent. In the following section will be examined more closely the making processes.

Ideas about making

> *"The person who recognises their own product in the world that has actually been transformed by their own work: he recognizes himself in it, he sees in it his own human reality, in it he discovers and reveals to others the objective reality of his humanity, of the abstract and purely subjective idea he has of himself." Alexandre Kojève (Crawford 2009, 14).*

Ada Nuottajarvi uses her research time at work to create new leather effects for wall panels or furniture. Some are speculative and their samples will be sent to the designers: if they like a certain effect they will integrate it into a design. This is how the corridor project would have started and it provides a vital opportunity within the work place to be creative: experimenting with leathers to produce new ideas about how to innovate. The designer may make modifications, specify a different colour or type of leather or make other adjustments to match the aesthetics of their design. This can be a straight-forward process in which the designer and maker work together. If a designer is working with many different materials it is not always possible to fully understand the properties of each material, thus advice from a maker may be critical. A maker will often work with a specific material for years, developing an innate knowledge of its properties and applications. Similarly, the designer has a valuable knowledge of the wider aesthetic of the project and of the client's requirements. If this process is clear and cooperative the question and choice 'Why leather?' can be comprehensively addressed.

Alternatively, the process can break down. This can lead to the specification of unsuitable leathers. The two best examples of misappropriated leather specification known to the author give little consideration of the material properties — not just of a particular type of leather, but of leather as a whole — resulting in hugely impractical applications. The first was the covering of plant pots in fine goatskin leather; the second was leather floor tiles in a designer shoe shop in which many women would be wearing stiletto heels. The question 'Why leather?' in this context becomes superficial because similarly the material (leather) is considered on no more than a perfunctory basis.

There is a current trend in some aspects of education and industry that facilitates a division in the processes needed for successful making and design. In this regard fine art is self-contained, it comments, reflecting back to the observer different perspectives on the world. It is a communicator of ideas, not a functional object. Design is primarily functional but often made more desirable by aesthetic success.

Skilled making often crosses into the disciplines of both art and design but is often maligned as decorative substitute for either. Design is seen as more intellectual than making, but has the danger of becoming abstract without the grounding of materiality. The realization of abstract ideas can often fall to the maker, meaning they also assume the role of designer. Contemporary training often puts emphasis on one aspect, negating the other; a balance between the making process and creative thinking has the most potential to create interesting work. As expressed in her interview, Caroline Grappe believes the lack of continuity in the teaching of these skills is where the problem often begins. Talking of her experience of training on bookbinding and book arts degrees:

> *"It is worrying that the emphasis on design usurps the need to learn technical and material skills. It seems that this starts in education with lack of technical learning but I have also encountered it in industry with the proliferation of designers whose practice has no basis in making or a maker's knowledge of materiality. I am concerned about the loss of skills which the continuation of these trends will cause."*

Without the technical skills learnt on work experience and her own propensity to make, she felt she would have struggled to make a technically proficient book on the completion of the course. Ada Nuottajarvi, however, experienced a more integrated educational approach in Finland:

> *"The assumption was that students on my course would become teachers, not as a diversion from their own practice of the subject, but in the many arts and crafts centres that still exist around Finland. Here traditional skills are taught to whoever wants to learn them, the things that are made and sold utilise these skills in traditional and contemporary ways but a culture of making is continued."*

Ada Nuottajarvi's course taught technical skills in a historical context and design skills rooted in material knowledge. The design process would start subsequent to properly understanding the material. This is a crucially important process, which needs to be echoed in the workshop. Ultimately, even at the high end of the market, economics rule, but we have to be careful not to lose valuable skills and knowledge in pursuit of commercial expedients.

The different factors which contribute to the use of leather as a decorative finish in the luxury interiors industry have been examined and many of the making and production processes involved in a particular material: vellum. The role of the maker as a practitioner of hand-worked, traditional techniques entails an involved process of learning and practice. This process can facilitate an in-depth knowledge of leather enabling the maker to transform the ways in which it can be used. This in turn provides innovation in leather use, which given cooperative working practices, can be utilized by the designer and presented as exclusive to the client. The patronage of the client enables the maker the time to continue to innovate with leather.

In this cyclic process each party has their own answers to the question 'Why leather?' In summary, the maker appreciates a diverse material with the potential to develop and innovate, the designer thinks about leather in the wider context of their design and the client appreciates the fashion, status or aesthetic qualities leather might bring to their interiors. Project Two involves less skill on the part of the maker, and examines some of the processes which occur when moving away from exclusively handcrafted making.

Project Two: the shagreen table

The synthesis of the handmade, machine made and contemporary technologies

Mapping out the project

A piece such as the shagreen table renders the maker purely technician: the method is prescribed, yet the beauty of the finished piece makes its execution satisfying. This table was designed and made by a bespoke furniture maker. The quality of the carpentry is good and it came with technical drawings made by the furniture company. Next, the tables needed to be covered in shagreen skins (figure 8). The colour of the shagreen skins had already been negotiated with the designer. Once the table was measured out and the number of skins ascertained to be correct, laser cutting of the skins could start. The use of technology means the maker has to use less skill and ingenuity in manipulating the material; a general aptitude for this kind of work would be sufficient to execute this project.

Why shagreen?

In this instance the maker is working with the shagreen in a less complex process than was previously demonstrated with vellum in Project One. However, the end result is equally impressive. In an expedient working process, the pieces are glued quickly because the maker is working with clean machine-made cuts, there is no messy manipulation of the leather or awkward joins to contend with. The glossy, opulent surfaces that are created are currently a popular design aesthetic. To a client, or even designer, it may not be recognised but the maker may feel the making processes involved have been curtailed. Before we examine this further we will look at the production processes used in both curing and working with shagreen.

The material: shagreen

Shagreen is the skin of the stingray *(Dasyatis bleekeri)*. It is covered with round, closely set calcified papillae, which resemble small pearls with one predominant pearl in the centre of the skin. The skin is cured rather than tanned. It is often

Figure 8. Detail of the shagreen table. Courtesy/copyright of Harcourt Design Ltd. Photograph by Harcourt Design Ltd.

polished down to create a shiny surface with a prominent pearl and dyed from the reverse side infusing the epidermis with colour and highlighting the pattern of the pearls. Its use in the decoration of interior objects has a legacy; through the inlay work of Jacques Galuchat shagreen was popular as a decorative feature of furniture in the 17th century court of Louis XV and it had a renaissance with art deco design of the 1920s and 1930s (Brunn 2011).

Contemporary making processes

Contact adhesives

Contact adhesives can range from extremely expedient (and toxic) spray glues to roll on/brush on white glues and neoprene. Glue is applied to both surfaces which bond on contact. This gives less manoeuvrability than wet glues but much faster drying times. The calcified structure of shagreen means it cannot be moulded around a surface in the same way as vellum or leather and therefore the glue does not have to penetrate the surface and make the shagreen malleable. Contact adhesives stay on the surface of the material so the shagreen is not stretched and consideration of shrinkage is not fundamental to the gluing process.

Laser cutting

The calcified papillae in the structure of the shagreen skin make it very difficult to cut straight precise lines. It is possible to cut with a scalpel or Stanley knife but depending on the area of the skin results can vary sometimes leaving an uneven edge. The edges (belly and head/tail of the skin) tend to have a finer structure of papillae and a lesser thickness of skin and are therefore easier to cut accurately (figure 9). The laser cutter can accurately cut through any thickness of shagreen skin, creating precise edges. This is particularly important for this project as the shagreen pieces are butted up against each other rather than overlapped at the join (figure 8).

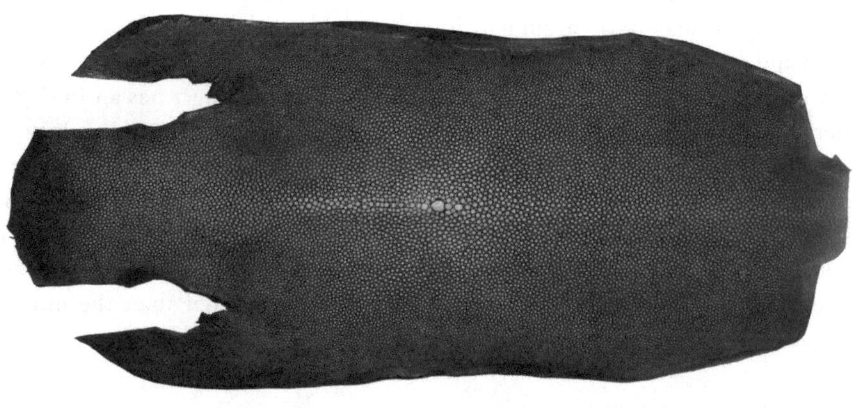

*Figure 9. Shagreen skin (*Dasyatis bleekeri*). Photograph by Laura Youngson Coll.*

Problems with shagreen

Shagreen is very popular as a decorative finish for luxury furniture and goods. It carries a certain status as an exotic material, and its shiny polished surface can make its aesthetic quite opulent. However, the source of skins is often Thailand and information about their origin is scarce. Before the trend for shagreen as a luxury material re-emerged in the 1980s shagreen skin was a by-product of Indo-Pacific fishing. As its popularity has caused an industry for the skin alone and the meat is of relatively low value, farming of stingray for skins is common. The Convention on International Trade in Endangered Species of Wild Fauna and Flora (CITES) certification is not given to this species, and farming prevents depletion of the species in the wild, but there also seems to be little information on the farming methods. This has to be taken as an ethical consideration in the use of shagreen. The other consideration is the waste of skins, which is a consequence of the central large 'pearl' or 'eye'. This feature often appears in the middle, or slightly off centre of a piece, which forms part of a geometric pattern. A small rectangle is taken from the centre of the skin leaving a lot of waste. This can be used in smaller items such as door handles, or through a design which utilises the whole skin.

Ideas about contemporary making

In the context of bespoke leatherwork, combining handmade work with computer led technologies can produce varying results. Some of the laser cut patterns and etchings are remarkable and can be combined with other leatherworking techniques to produce beautiful designs. However, there are limitations on how the laser machine can engage with the leather. It can cut intricate patterns and, on a basic level, is set to understand the material properties of different leathers but it cannot pare, gouge, shape or construct: essentially it cannot interact with the leather. The argument can be made that the creative response is made in the act of drawing the design, which the machine then follows. To some extent this is true but it negates a direct response to the material in question: the leather.

This process may be an argument for the considered but combined efforts of human and machine, much of the basis of contemporary making but the question 'Why leather?' is asked on a different basis. The maker no longer has an in-depth material knowledge of the leather to pass on to the designer, although this may be replaced with knowledge of how the leather reacts to more technological processes. This shift in making technologies probably has more effect on the maker than the designer or client, although easier production methods make the craftsmanship of the material less exclusive. Faster and easier production methods could facilitate cost efficiency but ultimately becomes mass production rather than the unique product of a creative process.

Slick surfaces and fast living

The opulent, almost glitzy, quality of the shagreen is indicative of its current popularity as an interior material. Some of the interiors which we encounter in our work are at the beginning of their life. More often than not, it is the designer and not the owner, who is overseeing the work. This creates a strange sterility in many of these environments. Accentuated by slick shiny surfaces and lack of personal effects, it is sometimes difficult to imagine these interiors functioning as a home.

> *"The entire concept of decoration has changed too. Traditional good taste, which decided what was beautiful on the basis of secret affinities, no longer play a part here. The taste constituted a poetic discourse, an evocation of self-contained objects that respond to one another; today objects do not respond to one another, they communicate-they have no individual presence but merely, at best an overall coherence attained by virtue of their simplification as components of a code and the way their relationships are calculated." (Baudrillard 1996, 23)*

The unease felt when encountering these environments is perhaps a reflection of a wider phenomenon of 21st century living: the trend for an expedient, ever changing style of living combined with the superficial nature of status driven objects. Even in the high end interiors industry this attitude can facilitate a throw away culture. I have been given accounts of projects in which whole interiors have been completed to an extremely high specification, sold, ripped out and the whole process is started again. This obscene waste of materials, labour and resources also encourages disrespect for craftsmanship. If it is likely that something will be replaced in five years' time why consider the longevity of the material or the long term potential of the production involved. If less time is invested in the production process, the immediacy of the product becomes more economically viable, but the legacy of both craft and environment ultimately becomes depleted.

The shagreen table presents a contrasting piece to the vellum ribs. Although both leathers are currently fashionable, and therefore are favoured in contemporary interior design, there are factors which differentiate the two pieces in their provenance and making processes. The more expedient production methods of the shagreen table seem to be reflected in its showy aesthetic and lack of concern for the provenance of the skins. However, the laborious production time and subsequent cost of the vellum pieces mean they can only be produced with substantial patronage. It could be argued that this makes them equally exclusive and showy, even if through a subtler aesthetic. The question 'Why leather?' becomes multifaceted because the materials defined as leather, and the working methods employed with interior leathers, are also multifarious. Before we discuss this further we will briefly look at one more material used in a different interiors context, vegetable tanned leather.

Project Three: vegetable tanned goatskin leather

The maker's investigation of leather and vellum

Vegetable tanned goatskin is used to make fine decorative leatherwork. The leather is tanned using bark and tannins from plant sources and dyed in a diverse range of colours. It is commonly used in bookbinding to create intricate cover designs. It is skived,[1] pared or thinned to create extremely fine pieces that are used as on-lays to produce decorative effects. The decorative panels shown (figure 10) refer to these techniques combined with an appropriation of paring techniques.

These pieces are made on a non-commercial basis giving more scope for exploring the material and aesthetic qualities of various leathers without the restrictions of time or client specifications. Decorative artworks, made from tiny pared pieces of vegetable tanned goatskin and vellum, exploit the grain and textural qualities of the material. The maker does not have to create a theoretical brief because ideas are actualised as a direct result of the making process. All the skills and knowledge learnt through repetitive making and experience are utilised to exercise the material and aesthetic possibilities of the leather.

Similarly, this integral approach to design and making is expressed in Ada Nuottajarvi's experimental designs (figure 11). Ada describes her approach to leather:

> *"I see leather in a similar way to any number of other fabrics; each with their own properties and material demands. When I first started working with leather, as opposed to many other fabrics, one of the material properties I really appreciated was non-fraying edges."*

Making facilitating art

The pieces shown in figure 10 are intended as decorative artworks for interiors, and are made with this in mind. So although they are more experimental than commission based work there are still some practical and aesthetic restrictions placed upon them. Figure 12 shows the initial stages of work that creates a conceptual response to our interaction with the environments in which we live.

Discussion

Three different leathers, vellum, shagreen and vegetable tanned leather, and their use in contemporary interior leatherwork has been discussed. The question 'Why leather?' is applicable to each material based on a number of differing factors: the tanning, curing and procurement methods, the different making skills and techniques required, the dialogue between maker and designer. The perspective of each of the parties involved in the production process is also of importance, including the maker, the designer and the client.

In Project One the maker chooses to work with vellum because potential is recognised as unusual material, contributing to the appreciation and aesthetic qualities. Through experimentation, and the adaptation of traditional leatherworking techniques, innovative work is produced. Of vital importance to the maker is a direct, interactive response to the vellum; a material which has such potential is an exciting prospect to a maker who has trained to consider and exploit

1 Skiving is a mechanical method of paring using a skiver, a traditional leatherwork machine. The leather is fed through a blade which slices into the leather; it results in a more linear but less adaptable method of thinning leather than hand paring.

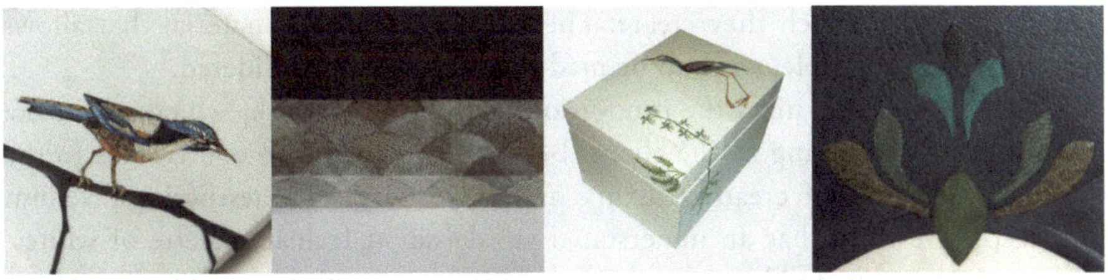

Figure 10. Decorative leatherwork. From left to right: Garrulinae (detail), vegetable tanned leather on vellum; Mesoderm mirror (detail) 'Supernatural' vellum; Year One box, vegetable tanned leather on vellum; Trifoliate (detail) vegetable tanned leather. Designed/made and photographs by Laura Youngson Coll.

Figure 11. From left to right: Leather cushion with moulded pattern; sewn leather necklace. Designed/made by Ada Nuottajarvi. Photographs courtesy of Ada Nuottajarvi.

Figure 12. Geodesic Lichen Sketch-vegetable tanned leathers; Lichen Kepler-vegetable tanned leathers, wire. Designed/made and photographed by Laura Youngson Coll.

the material with which they create. The value of finding a material that allows beautiful, and profitable, work to be made should also be considered.

The designer recognizes the innovation in the vellum work, which will make it unique and interesting to the client. Bespoke craftsmanship contributes to the designers' objective of creating luxury and exclusivity. The aesthetic of vellum as a decorative finish has an understated yet decadent feel: its palette of whites, creams, muted greys and browns and joins which create subtle patterns. The client appreciates vellum as an aesthetic or status material dictated by fashion or personal taste. The relative rarity of vellum and the skilled making involved in its production may contribute to this appreciation. Although vellum is durable, unprotected it can become scuffed or damaged and, moreover, humidity can lead to distortion. The vellum may be lacquered to remedy this problem but the client does not necessarily have to consider these practical implications. Once the vellum panels or furniture are installed, environmental conditions are regulated and specialist cleaners employed, further emphasizing the rarity of the material.

Project Two is a collaboration between maker, computer and laser cutting machine. Shagreen is recognised by the maker as an opulent material that fits well with the current aesthetic in interior design. It is also recognised as a material which is difficult to cut precisely by hand. The maker has to strike a balance between machine and hand work to create the desired effect. The more work the machine can make, the easier the production is for the maker but the closer the object becomes to being manufactured, rather than handcrafted. The maker may feel a lack of experimentation and making process makes them a low skilled technician when working with shagreen in this way. The material is used to compensate for the making process as the glitzy aesthetic and high value of shagreen skins ensures their popularity as a status material. The designer may also respond to fashion in this way using shagreen to create an interior dripping with opulence. The provenance of the shagreen skins should cause concern to all involved but unfortunately scarcity can also lead to exclusivity, a quality often considered highly by designers and clients alike in luxury interiors. The processes and materials in Project Two have a superficial taint, from expedient making processes to an aesthetic of blatant glitz, but this does not detract from the success of shagreen as a luxury interior material.

Project Three examines briefly the application of a maker's knowledge to projects outside the remit of the designer's brief. Here the maker chooses fine vegetable tanned leathers as they provide a palette and texture that allows the creation of image and pattern. The maker sees the leather in the same arena as other materials and textiles, assessing and working with its qualities in a direct response, and then reconstructing the results. Small pieces of leather are recycled exploring more economic ways of utilizing the resource. More freedom is afforded in the making process but as the work is not made to a designer's brief this work is purely speculative. The designer or client may recognize the work as uniquely crafted and of the right aesthetic but there is no guarantee of making the work commercially viable.

Conclusions

The question 'Why leather?' is addressed within a contemporary but niche area of design. This is not a utilitarian application of leathers but a highly specialized part of the luxury design industry. This is indicative of leather, especially unusual leathers such as those examined here, being used as a signifier of luxury. Vellum or shagreen can be chosen by a designer to embellish a lavish interior with an expensive and bespoke material but the question 'Why leather?' is more nuanced for the maker. The specific leather, its properties, aesthetic and value all dictate how the maker responds and creates with the leather. On a more practical level, time, and therefore money, has to accommodate such a process. While experimentation, and consequentially innovation, cannot always be directly turned into a tangible product, it is fundamental to the reason why a designer, or client, would choose the work of such a maker.

The process which poses the question 'Why leather?' in contemporary interior design becomes cyclic in nature. In summary the examined leathers have status or material interest because they are unusual, rare or aesthetically pleasing; this is intrinsically linked with the maker using them in innovative ways. This is facilitated by extensive training and learning which is ultimately funded by producing a specialist product which is further buoyed by an exclusive market. The niche for craftspeople and specialist leatherwork in the 21st century may not be so different from those of centuries past.

References

Baudrillard, J. (1996). *The system of objects.* Trans. Benedict. J. London/New York: Verso.

Brunn, M. (2011). Provincial Museum of Alberta. Conservation DistList Instance. http://cool.conservation-us.org/byform/mailing lists/cdl/2001/0513.html. [Accessed: 16 August 2011.]

Crawford, M. (2009). *The case for working with your hands or why office work is bad for us and fixing things feels good.* London: Penguin Books.

Lucie-Smith, E. (1979). *Furniture: a concise history.* London: Thames and Hudson.

Rix, M. Sherwood, S. (2008). *Treasures of botanical art.* London: Kew Publishing.

Skelton, I. (2010). Die Frau Metta Catharina von Flensburg: a Danish brigantine wrecked in 1786 in Plymouth Sound, England. *International Journal of Nautical Archaeology* **39** (2). 235-257.

Museum exhibits

Victoria and Albert Museum: Exhibit museum no. P.26-1975. Viewed on 21 June 2011.

Interviews

Grappe, C. Bookbinder and leatherworker. Interviewed 25/08/2011.

Nuottajarvi, A. Bookbinder, leatherworker and jewellery designer. Interviewed 25/08/2011.

Acknowledgements

Thank you to Dr. Susanna Harris for inviting me to embark on this paper. I am grateful to Ada Nuottajarvi and Caroline Grappe for their time, knowledge and generous giving of images and description of their work. Thank you to Gavin Rookledge of Rook's Books Ltd, and Simon Harcourt and Emily Cramp of Harcourt Leather Design Ltd, for permission to reproduce images of work. Thanks too to Richard Sharpe for patience, support and proof reading.

Leather in the textile industry

A memoir

Alan S. Raistrick

Preface

I was brought up in a family of tanners making leathers for use on machinery mostly in the worsted textile industry around Bradford. I remember my first visit to the tannery when I was about five years old and my first visit to a textile mill was when I was seven. Most of the information in this paper was learned at my father's knee and gave me the training by Patrimony which would have gained me automatic entry to a medieval gild. Just like the trade secrets of a medieval gild, much of the contents of this paper have never been published before, so far as I know. When I was doing research for the leather industry, from 1958 onwards, I realised that most of the leather industry was not aware of this highly specialised branch of their trade.

Introduction

With both wool and cotton the raw material is in a random mass and needs organising into a continuous ribbon in order to feed the spinning machines. This used to be done by the hand spinners' fingers, but the machine inventors had to somehow do it by machinery. The strips of fibre, either ready for spinning or partway processed, were variously known in different branches of the textile industry as 'sliver', 'slubbing' or 'roving'.

Almost every machine in the textile industry relied on leather in order to work. General knowledge of leather in the textile industry is restricted to the enormous forest of drive belts that festooned the factories, which did form a large part of the production of the tanneries. For one or two machines the drive belts were the total involvement of leather, but most relied on leather to grip, pull, and support the fibres in processing.

The different animal skins available give a range of thicknesses from a millimetre or two with sheepskins up to around twenty five millimetres for bull shoulders, with the average cattle hide around five to eight millimetres. The tanner can adjust the properties from very flexible to very rigid. The surface properties can be made such that the leather can grip the fibres being processed but they do not stick to the surface. The development of machines using leather and replacing hand processing steadily increased from 1769 to around 1850. The decline in use of leather started around 1960, to be replaced steadily by plastics and is now complete.

The machines

Arkwright's spinning machine

The first usable spinning machine alternative to the spinning wheel was Richard Arkwright's patented machine of 1769 (English Patent No. 931,1769, Richard Arkwright) (figure 1). For the drafting rollers of his spinning machine of 1769 Richard Arkwright after much difficulty found that a roller covered with alum tawed sheepskin pressing against a fluted wooden roller gave sufficient grip to draw out the sliver of fibres before spinning on what amounted to a standard bobbin/ flyer spinning wheel on its side. Industrial archaeologists endlessly refer to this as his 'water frame' but it is not: it was totally unsuitable for continuous spinning because it had to be stopped frequently to change the thread on to the next hook on the flyer.

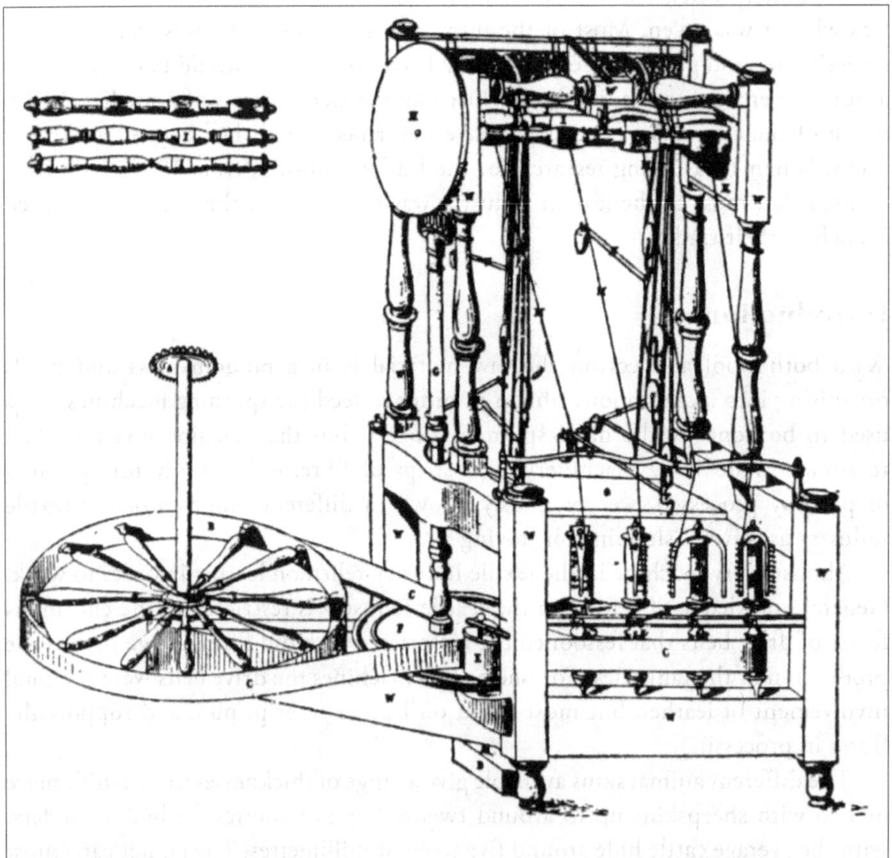

Figure 1. Arkwright's Patent spinning machine (English Patent No. 931,1769, R. Arkwright, 1769).

The throstle

Shown in figure 2 is a version of the water frame called a 'throstle'. The hooks on the flyer have been removed and the frame carrying the bobbins now moves up and down inside the flyer, to give continuous wind on. The drafting rollers are now covered with vegetable tanned sheepskin leather. The brake bands on the bobbins have been found to be unnecessary and have been removed. The water frame at first proved to be a mixed blessing, because it still had to be fed with a sliver prepared by a hand spinner on a spinning wheel. This machine could not be patented because the law at that time did not allow improvements to patents to be made. Arkwright had to develop a whole system of machines to keep his spinning machine supplied with prepared fibres (English Patent no. 1111, 1775, Richard Arkwright).

Hand carding

Hand cards (figure 3) have been used for preparing wool for spinning since at least the 14th century AD. They are illustrated in the Luttrel Psalter of ca. 1340 (British Library Add 42130, f193) and the roughly contemporary Smithfield Decretals (British Library Royal 10 E. IV f147v) both of which clearly show the bent wire staples. Later they were used for preparing cotton. The leather backing has been found in archaeological digs from a similar period.

The carding engine

Arkwright used similar card clothing with staples inserted through a leather backing on his carding machine (figure 4), as can be seen on the machine at Quarry Bank Mill. The teeth are only brushed past each other, not pressed together with the teeth between each other. He covered drums with card clothing which he rotated close to each other to open up the cotton or fleece and reorganise the fibres. For a woollen yarn this is into a random arrangement. Around a hundred years ago the leather backing was replaced by heavy rubberised fabric.

Tape condenser

The carding engine consists of many rollers working against each other, finally producing a wide web of fibres. This particular machine uses chains for driving all the rollers. On the right hand side it can be seen that the machine is producing a set of slivers ready for spinning. The part of the machine doing this is known as the 'tape condenser' (figure 5).

Worsted card

This carding engine (figure 6) is for worsted yarn, where the fibres are organised parallel to each other. Notice that this machine happens to have leather belts driving all the rollers. This appears to be simply at the whim of the machine designer.

Figure 2. Throstle. Courtesy of the Helmshore Textile Museum. Photograph by Alan S. Raistrick.

Figure 3. Present day hand carding. Photograph by Alan S. Raistrick.

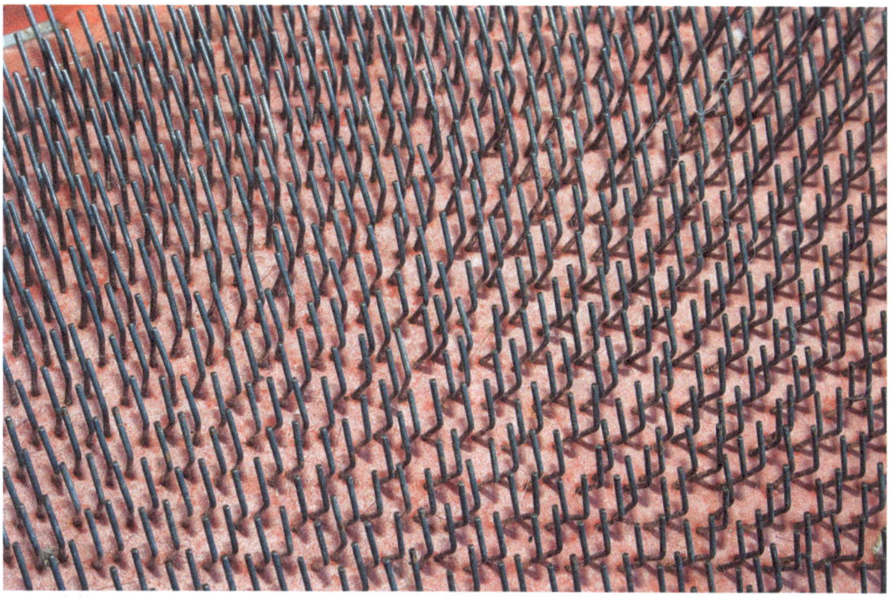

Figure 4. Modern Card Clothing.
Photograph by Alan S. Raistrick.

Figure 5. Tape Condenser. Courtesy of the Saddleworth Museum.
Photograph by Alan S. Raistrick.

Ring doffer and rubber

The carding engine shown in figure 7 has a final roller covered with strips of card clothing to pull separate strips of fibres off the previous roller. Known as a 'ring doffer' it oscillates gently from side to side as it rotates, thus removing all the web and not leaving strips of fibre undoffed. The flat strips are then passed through a pair of wide short leather belts which feed the sliver forward and at the same time oscillate from side to side rolling the flat strip into a round sliver ready for further processing as needed. The leather for this has to be very flexible because the rollers it goes round are of small diameter and it needs to have a very smooth surface free of faults. Traditionally the Swiss farmers cared for their cattle better than anyone else in Europe, so Swiss hides were preferred. The rubbing leathers have to have a joint which is adequately flexible. Cellulose nitrate cement became the preferred glue.

Tape condenser and rubber

Alternatively a tape condenser could be used to divide the carded web into strips (figure 8). The tape can be seen followed again by a rubbing leather or rubber to roll up the flat strips. The tape may be a single one up to 1,100 feet long, which works its way across the machine and is then fed back underneath from the last side to the first. The tapes may be either of leather or steel. Steel has a tendency to cut the fibres whereas leather pulls them apart less brutally. If a break occurs rethreading is a nightmare. Separate tapes may be used, but may be a nuisance if differential stretching happens.

Condenser tape diagram

The web that comes off the carder is fed into the tapes which have a path rather like a pair of scissors (figure 9). The tape from the top roller pulls a strip of the web down, while the tape coming off the bottom roller pulls the strip of web up. Joins in the tapes present a serious problem.

Wool combs

The alternative method of preparing fibres is by combing (figure 10). Wool combs have been known from Viking days (Malik 2012, 85-130), which tended to have a single row of tines rather than the multiple rows on the illustrated pair. The combs were heated in a charcoal or coke heated pot. The industrial revolution reached this trade between 1851 and 1861 as shown by the census records of those years. Combing was for the longer fibred wools, to straighten the fibres, and make them parallel. The long, combed strands were known as 'tops'. The short fibres separated in the combing were known as 'noils' and were sold to the woollen trade. The tops were then spun into worsted yarn. Woollen yarn had a random arrangement of short fibres.

Figure 6. Carding engine. Courtesy of the Leeds Industrial Museum. Photograph by Alan S. Raistrick.

Figure 7. Ring doffer followed by a rubbing leather. Courtesy of the Saddleworth Museum. Photograph by Alan S. Raistrick.

Figure 8. Tape condenser followed by a rubbing leather. Courtesy of the Saddleworth Museum. Photograph by Alan S. Raistrick.

► *Figure 9. Condenser tape diagram. After: Spibey (1952).*

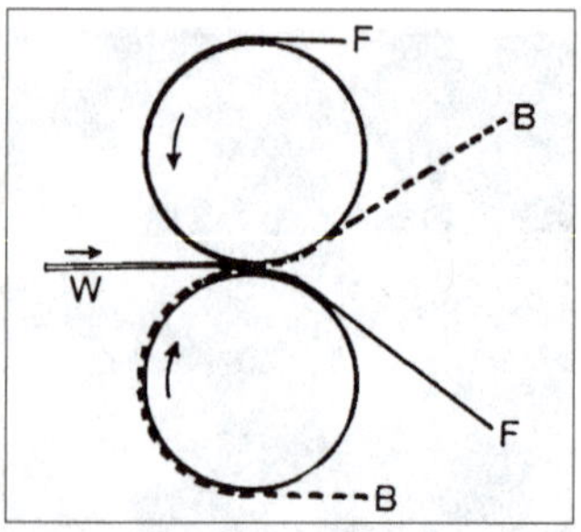

Figure 10. Wool Combs. Scale bar represents 12 inch. Photograph by Alan S. Raistrick.

Gillbox feed

In the middle of the 19th century AD it became clear that the two processes in hand combing had to be separated if it was to be done mechanically. The gillbox did the straightening and the comb did the separation of the short fibres from the long fibres. The name 'gill box' appears to be from a corruption of the French '*aiguilles*' for needles. The gillbox also served to produce an even yarn by mixing two, or more, slubbings or slivers and drawing them together to form a more even yarn.

Gill box

The slivers were pulled through the first rollers by sets of moving pins, which can just be seen in the middle of the photograph (figure 11). The combing leathers to the left, running over corrugated rollers that leave their mark on the leathers, then pull the wool sideways off the pins. A gillbox might take two slivers and with a draft of four, produce a slubbing half the size of each feed slubbing. Repeated gilling steadily produced a finer and more even slubbing. A bad joint in the leather could produce a cyclical variation in the slubbing. The leathers were by the 1950s chrome tanned, but if a mill manager preferred the old fashioned brown kind of leather he remembered from his youth, he was duly supplied with it by swabbing the blue-green chrome leather with a spirit based brown dye.

Preparer feed

The first gillbox in a set was known as the preparer gillbox, which had a leather feed apron. Figure 12 is included as it shows one way of joining leathers where the joint does not affect the gripping performance. The joint can be laced rather than glued. The laces were made from sulphur and oil tanned E.I.Kips. A 'kip' is a skin half way between a calf and a full grown bovine. Indian cattle with a hump have a thinner skin and are smaller than European cattle and a full grown skin is kip size. 'E.I.' is short for 'East Indian'. While leather can be split, the strength of each split piece is less than half the strength of the original leather, so for lace leathers, where strength is important, it is essential to use the whole thickness. The tannage consists of soaking the skin in sodium thiosulphate solution followed by acidification to precipitate sulphur inside the skin. Then treating with cod liver oil produces a thin, white, strong, flexible leather suitable for laces. The laced joint is useful for cases where leather stretch could be a problem as it is easily remade by cutting off a strip of leather from the belt and re-lacing the joint.

Lister comb

Attempts to produce a combing machine to separate the long fibres from the short were made throughout the first half of the 19th century without success. The Lister comb (figure 13) was the first successful machine, devised by Samuel Cunliffe Lister, appearing in the middle of the century. By 1861 the hand wool combers in my home village of Oxenhope, West Yorkshire, had virtually disappeared. There is

Figure 11. Gill box. Courtesy of the Bradford Industrial Museum. Photograph by Alan S. Raistrick.

Figure 12. Preparer feed. Courtesy of the Bradford Industrial Museum. Photograph by Alan S. Raistrick.

Figure 13. Lister comb. Courtesy of the Bradford Industrial Museum. Photograph by Alan S. Raistrick.

a circular comb on the left of the machine onto which tufts of slubbing are placed. Then the longer fibres are pulled off by a pair of leathers running over rollers to form a new slubbing fed out at the front. The comb moves on and the short fibres are removed; the comb moves around to be refilled continuously. In operation it has a wonderful Heath Robinson appearance.

Noble comb

The Lister comb was quite rapidly followed by this Noble comb (figure 14), again with the involvement of Samuel Cunliffe Lister. It is much neater, but has to be fed by discontinuous balls of slubbing. Peering down the top of the Noble comb (figure 15) you can just about see that there are a large ring comb and two smaller ring combs working with it. There are pairs of leathers set vertically to pull the long fibres out of the combs, in all four pairs. The combs are heated just like the hand combs to help straighten the fibres.

Halifax flyer frame

To draw down a slubbing to a finer state a draw frame could be used. This is a version of Arkwright's machine. It can just be seen that the large rollers are covered with calf leather which is laced around the rim (figure 16). I do not have a photograph of the final spinning frame, but basically it used Arkwright's method derived from the bobbin/flyer spinning wheel.

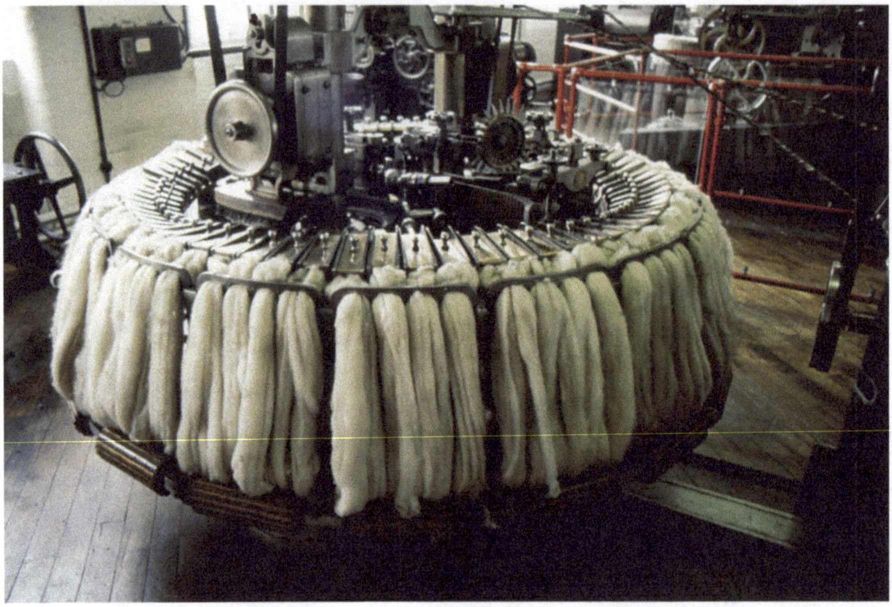

Figure 14. Noble comb. Courtesy of the Bradford Industrial Museum. Photograph by Alan S. Raistrick.

Figure 15. Noble comb with combing leathers just visible. Courtesy of the Bradford Industrial Museum. Photograph by Alan S. Raistrick.

Figure 16. Halifax flyer frame. Courtesy of the Halifax Industrial Museum. Photograph by Alan S. Raistrick.

Worsted mule

The alternative method of spinning the final thread, the 'mule', was derived from the spindle wheel shown in the Luttrel Psalter. James Hargreaves had produced his Spinning Jenny in 1764 shortly before Arkwright's patent. He took out a patent in 1770 but quickly lost it because he had not kept his invention secret before making his patent application. Samuel Crompton then developed his mule around 1780 which applied Arkwright's drafting rollers to the Jenny, which he could safely reveal in 1785 after the court case where Arkwright failed to renew his patents. The mule rapidly developed into the major machine for cotton spinning, eventually reaching a total of somewhere between 50,000,000 and 60,000,000 spindles, with each spindle needing leather covered rollers for the drafting process. It was found that the best leather, vegetable tanned, was made from Welsh Mountain sheepskins, and there was a string of tanneries from Machynlleth, Dolgellau, Newtown, Wrexham to Oldham where the biggest machinery maker Platt Brothers existed. I suspect that the tannery in Littleborough, north-east of Manchester, might have used the local, now rare, breed, the Lonk, which lived on the Pennine moors under conditions very similar to the Welsh mountains.

It has sadly proved impossible to find a photograph of a genuine fine cotton mule because not one is available in museums. Helmshore has a fine set of condenser mules which have no drafting rollers. New Lanark has a woollen mule from the Galashiels/Selkirk area, which does not have drafting rollers. There was a rumour that the Science Museum has a proper mule in store at Wroughton Aerodrome, but there was no reply to enquiries. In 1911 Platt Brothers made a pair of demonstration mules for Bolton Technical College and Bradford Technical College. The Bolton one is in store but can be seen by appointment. This Bradford mule (figure 17) is on display in Bradford Industrial Museum, and is actually a worsted mule. The mule was barely used for worsted spinning in Britain, but was heavily used in France, creating extremely serious competition to the Bradford trade. The row of leather-covered rollers can be seen in the close up (figure 18) and in the close up of the drafting rollers (figure 19).

Basic loom

As well as in the spinning processes leather was essential for weaving. Figure 20 shows an early power loom where the shafts are raised and lowered by leather straps running over rollers. This loom is also interesting because it is only capable of producing four sheds, and was described as a 'twill loom' in the early 19th century. Memory of this type of loom seems to have disappeared from history, although several hand loom examples from earlier years still exist in museums. Leather was frequently used in various kinds of shedding mechanisms. The skep in the foreground would have had buckled leather straps to keep the lid in place.

Over pick loom

On the 'over pick loom' the shuttle was knocked across the loom with a moulded rawhide picker made from buffalo hide (figure 21), pulled by a picking band, possibly again of buffalo hide, from the picking stick at the very top of the picture, possibly with the hair on. The shock of the movement would be taken up by bumper leathers, the white discs of formaldehyde tanned bull shoulder at the end

Figure 17. Worsted mule. Courtesy of the Bradford Industrial Museum. Photograph by Alan S. Raistrick.

Figure 18. Worsted mule drafting rollers. Courtesy of the Bradford Industrial Museum. Photograph by Alan S. Raistrick.

Figure 19. Close-up of the worsted mule drafting rollers. Courtesy of the Bradford Industrial Museum. Photograph by Alan S. Raistrick.

Figure 20. Basic early power loom. Courtesy of the Leeds Industrial Museum. Photograph by Alan S. Raistrick.

Figure 21. Over pick loom picking mechanism. Courtesy of the Bradford Industrial Museum. Photograph by Alan S. Raistrick.

of the shuttle box. At the back of the shuttle box, out of sight, is a brake known as a 'swell'. A swell may have been covered by a swell leather and caught the shuttle and stopped it bouncing back into the shed.

Under pick loom

The 'under pick loom' is so called because the picking stick is below the shuttle. Figure 22 shows the conglomeration of leather straps needed to control the mechanism. Here the bumper leather is in the form of a zigzag around the picker guide rod.

Picker

The picker illustrates one extreme of the wide range of materials that can be produced from hides. It (figure 23) is moulded from unhaired rawhide and dried to give a hard material resistant to the impact on the shuttle while knocking it across the loom.

Picking band

The picking band is strong and flexible, connecting the picker to the picking stick (figure 24). There was a belief in a kind of Samson complex: that it was stronger with the hair on. In fact it made no difference, but if the customer wanted the hair on, he got it with the hair on. It had to be flexible to wrap around the picking stick without cracking.

Figure 22. Under pick loom picking mechanism. Courtesy of the Bradford Industrial Museum. Photograph by Alan S. Raistrick.

Figure 23. Picker made of dried raw S.E.Asian buffalo hide. Photograph by Alan S. Raistrick.

Figure 24. Picking band with the hair on. Photograph by Alan S. Raistrick.

Shuttle

The shuttle in figure 25 has a fine synthetic thread carried on a pirn. The 'pirn' is a form of bobbin where the thread is wrapped around it when filling it, but the thread pulls off it over the end when weaving. If a bobbin is used there are problems due to the inertia when starting and stopping the bobbin rotating each time the shuttle moves. This shuttle is lined with wool-on sheepskin to control the thread as it unwraps off the end of the pirn. The thread tends to uncoil itself but the sheepskin gently holds it in place until needed. Larger shuttles for heavier threads may be lined with brushes to control the thread.

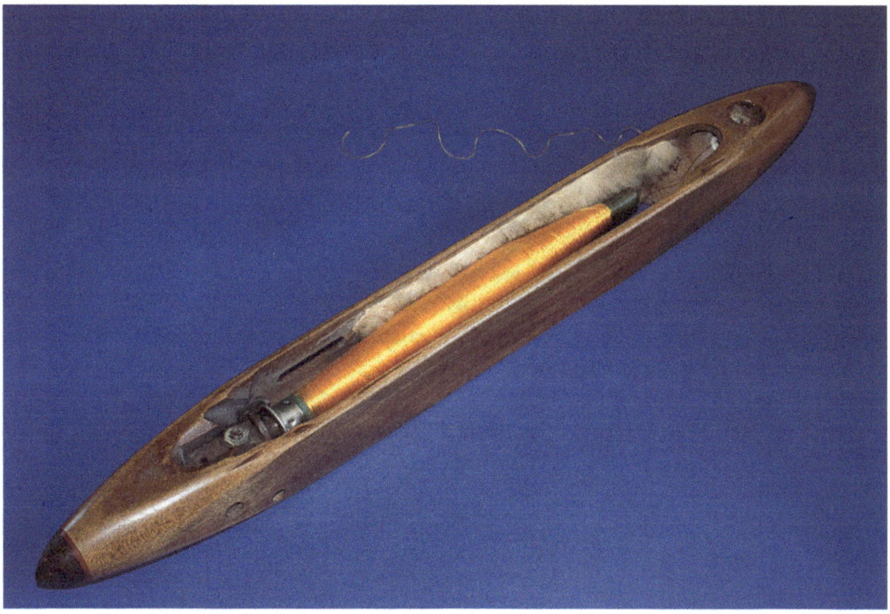

Figure 25. Shuttle with wool-on sheepskin lining. Photograph by Alan S. Raistrick.

Conclusions

I have shown that leather was vital for the development of the textile industry during the industrial revolution. The needs of the machines led to the development of a wide range of leather products, each particularly suited to the individual machines. The thickness range needed was accommodated by careful choice of raw material, from thin sheepskins through various thicknesses of different types of cattle, up to the extreme thickness of bull shoulders, developed by Mother Nature to protect the bulls in fighting. The types of tannage used varied widely, again selected to produce the characteristics for each particular purpose. This meant the Textile Leather Tanner used a far wider range of processes than for example a tanner only making shoe upper leather. The textile leather tanner was also unusual in the leather industry in that he did not sell leather, but sold the individual items needed by the textile machines, made up by the tanner from his leather.

Museums

Although most of the information here has not been generally published, most of it is readily visible in museums if you know where to look. It can involve putting your head inside a machine, preferably with the power turned off. Museums unfortunately tend to keep visitors at arm's length, well away from machinery. Interesting places to visit include Cold Harbour Mill, Uffculme, Devon; Bradford Industrial Museum, Yorkshire; Manchester Museum of Science and Industry; Quarry Bank Mill, Styal, Cheshire; Helmshore Mills Textile Museum, Helmshore, Lancashire; Saddleworth Museum, Saddleworth, Oldham; New Lanark Mill, Lanarkshire; (Welsh) National Woollen Museum, Drefach Felindre, Carmarthenshire; Leeds Industrial Museum,

Armley Mills, Leeds; Museum of Industrial Archaeology and Textiles (MIAT), Ghent, Belgium; and Textiel Museum, Tilburg, The Netherlands. This is far from a comprehensive list.

References

English Patent No. 931, 1769. Richard Arkwright.

English Patent No. 1111, 1775. Richard Arkwright.

Malik, K. (2012). The role of pit houses and gendered spaces on Viking-age farmhouses in Iceland. *Medieval Archaeology,* **56**. 85-130.

Spibey, H. (1952). *The British wool manual.* Buxton, Columbine Press.

Why leather in ancient Egyptian chariots?

André J. Veldmeijer & Salima Ikram

Introduction

Consulting a book sometimes leads to unexpected excitement. In Forbes' 1957 publication 'Studies in Ancient Technology' a photograph of a collection of horse trappings, housed in the Egyptian Museum, Cairo appears on p. 30, figure 6 (figure 1). The objects were not on display in the museum. However, when the photograph was shown to the responsible curator, Ibrahim el Gawad, he was quick to find the material in the storage magazines. The Ancient Egyptian Leatherwork Project (AELP) had already worked with el Gawad from 2008 onwards to study

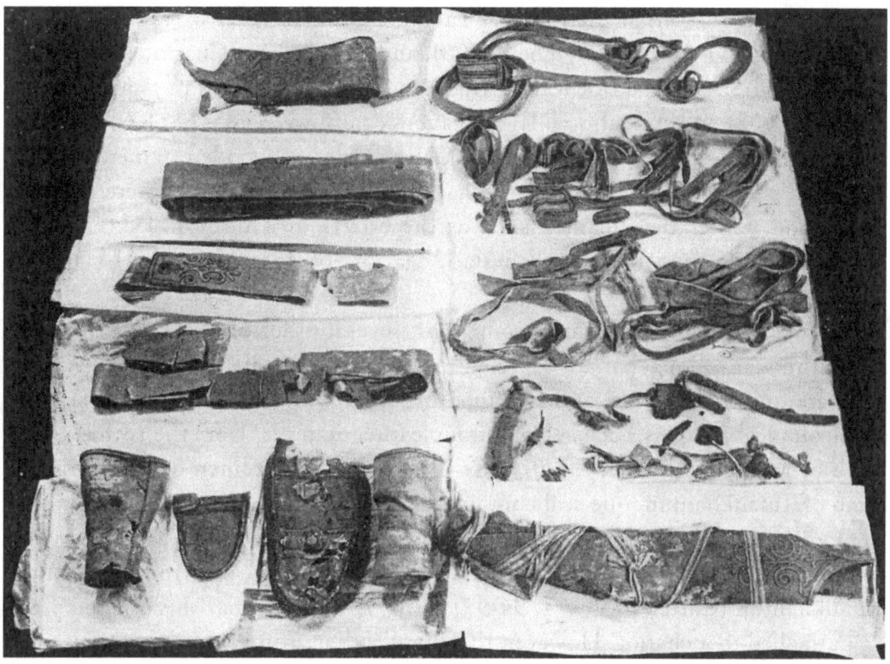

Figure 1. Part of the Tano chariot, as depicted in Forbes' 'Studies in Ancient Technology'. Courtesy of the MSA/Egyptian Museum Authorities.

the leatherwork held in the collection of the Egyptian Museum. However, the rediscovery of the chariot leather redirected the project's focus, resulting in the Ancient Egyptian Chariot Project (EMCP).

The find is nicknamed the 'Tano Chariot' because it was bought from a well-known dealer in antiquities, Georgios Nicolas Tano, probably in February of 1932. There is no further information about its provenance and date thus far; probably no records belonging to the dealer survive. However, a preliminary comparative study of the manufacturing technology suggests the leather is of New Kingdom date, probably late 18th or possibly early 19th Dynasty (Veldmeijer *et al.* 2013, 269).

The find is unique: the almost entire casing of an ancient Egyptian chariot, including parts of the bow-case that was attached to its side, as well as the harness. To the authors' knowledge, this is the only complete example of chariot leather that has been found. Thus, it has great potential in our understanding of not only chariot construction technology and use, but also, by extension, warfare and hunting, as well as the ancient Egyptian leather industry.

Egyptian chariots: a brief overview

A detailed description of the Tano chariot is forthcoming, but a brief excursus on Egyptian chariots is necessary here in order to be able to answer the question why leather was used for the casing and other parts. In ancient Egyptian chariots, wood was used for the frame of the box. Examples are the Tano chariot, the non-ceremonial chariots from the tomb of Tutankhamun (Littauer & Crouwel 1985), the Yuya and Tjuiu chariot (Quibell 1908), and the Florence Chariot (Botti 1951; Guidotti 2002). The chariot leather fragments from the tombs of Amenhotep II (Daressy 1902) and Amenhotep III (Littauer & Crouwel 1985, 68, 87), as well as the finds from Amarna (Veldmeijer 2010) also suggests a wooden framework only. However, there are also examples of chariots in which the frames were enclosed with wood, rather than leather, such as the box from Thutmose IV (Carter & Newberry 1904) and the so-called state chariots of Tutankhamun (Littauer & Crouwel 1985).

In ancient Egypt, the chariot came with several other objects, as is shown in the leatherwork shops that are depicted in two-dimensional art (figure 2). Among these are quivers that were attached to either side. Quivers are known from other archaeological contexts, such as the famous leather examples from the 18th Dynasty tomb of Maiherpri (Daressy 1902, 281-298) and the two linen quivers from the tomb of Tutankhamun, one still containing arrows.

Bow-cases, although often depicted hanging at the side of the chariot, are less common in the archaeological record. A wooden specimen came from the tomb of Tutankhamun (Carter 1933, 34, 94-97) but it is unlikely that this specimen was ever attached to a chariot. However, the Tano leather group includes various parts of a bow-case, to be attached to the chariot.

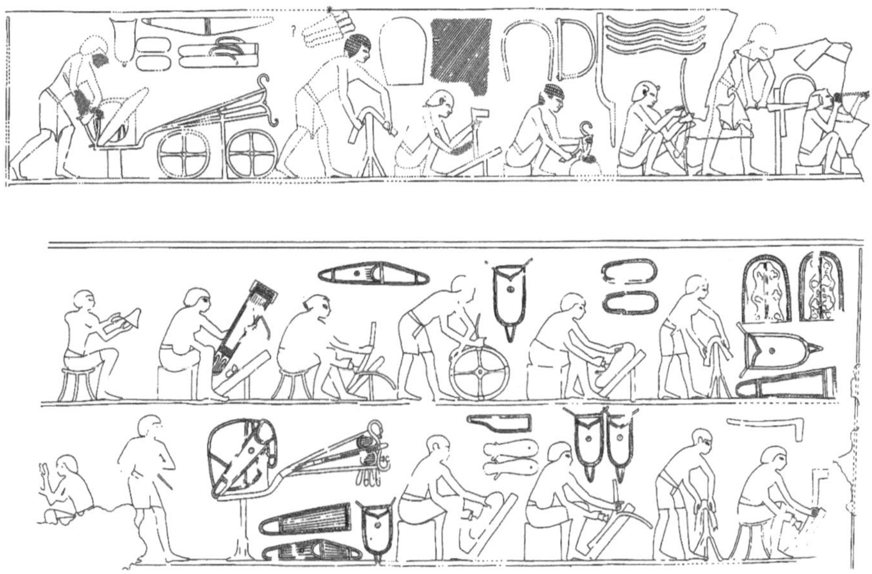

Figure 2. Leatherworkshop showing the manufacturing of various chariot related objects as well as the application of leather parts to the chariot itself, such as the tyres. From: Davies 1963, pl. VIII.

Bands used to secure the leather casing to the wooden structure are also a feature of chariots. The reins, harnesses, and horse trappings can also be included as part of chariot accessories, as can some of the accoutrements worn by the charioteer, such as wrist-guards.

The Tano leather

The leather fragments that constitute the Tano group include both large and small pieces (Veldmeijer & Ikram 2011, figure 3; Veldmeijer *et al.* 2013, figure 2). The pieces include the main portion of the casing as well as the straps and ties to connect the chariot together and to the horses. It appears that almost all of the leather portions of the chariot are present. Some of the pieces are decorated with leather appliqué work, while others are plainer. The leather was divided into two main groups, based on colour and robustness: red and green fine leather (figure 3), and beige and green robust leather (figure 4).

Choice of materials

Why was leather used in the Tano chariot, and what does that tell one about its use? The two so-called state chariots found in the tomb of Tutankhamun (numbered A1 and A2 by Carter; figure 5) are not cased with leather but rather are cased with very thin wooden board (elm in A2, Littauer & Crouwel 1985, 92; the wood in A1 was not identified, Littauer & Crouwel 1985, 93) that is covered with a layer of gesso and overlaid with gold. It is inlayed with glass and semi-precious stones. This decoration, together with constructional details, points to chariots that were

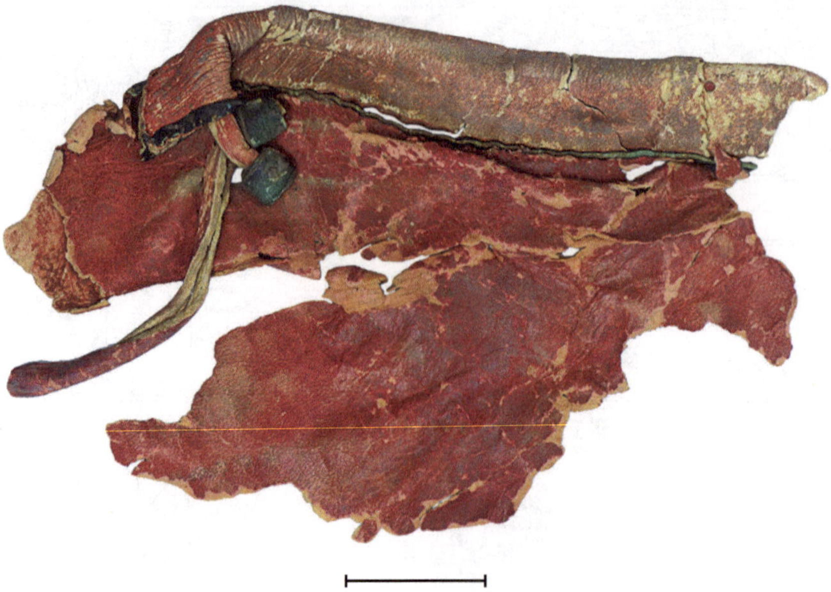

Figure 3. Large piece of the casing of the chariot's body. Scale bar is 5 cm. Photograph by A.J. Veldmeijer. Courtesy of the MSA/Egyptian Museum Authorities.

Figure 4. One of the two neckstraps. Scale bar is 5 cm. Photograph by A.J. Veldmeijer. Courtesy of the MSA/Egyptian Museum Authorities.

used for ceremonies or processions and the like, rather than on a battlefield or for other rough duties. The chariot from the tomb of Thutmose IV (CG 46097) is also made of wood and covered with gesso (Carter & Newberry 1904). However, texts (Littauer & Crouwel 1985, 99 and references therein) note that sumptuous decorated chariots were used by European kings to lead his troops in battle:

> *"such chariots might have been carefully brought along for the kings and princes to lead their forces in on the day of battle, even as the mediaeval knight's destrier was ridden only in "action"" [quotation marks in original].*

Figure 5. Overview of the Antechamber with the box of one of the State Chariots clearly visible. Photograph by Harry Burton. Copyright Griffith Institute, University of Oxford.

The thin board sidings would not have been strong or flexible enough for use in battle, and the gesso and gold would have easily fallen off in rough terrain. But can one distinguish between a ceremonial chariot and a war chariot? The above mentioned leather finds from the tombs of Amenhotep II and III that have been identified as portions of chariot leather clearly point to leather casings for chariots, unless these are the parts of the charioteer's gear and horses' harness rather than the chariot itself. If it is part of a casing, then does this point to a 'war or hunting chariot' rather than a 'ceremonial chariot'? Quite possibly: Thutmose IV, and especially Amenhotep II were great warriors and sportsmen and probably were active on the battlefield themselves.

Four chariots from the tomb of Tutankhamun had leather siding fill (A3-6, Littauer & Crouwel 1985). One had leather siding pierced by fenestrations (large openings in the casing of the chariot) and was nicely decorated, but was far less elaborate than the state chariots. Possibly the leather chariot fragments from the tombs of Amenhotep II (KV35) and III (KV[WV]22) came from such a chariot, an idea that has already been suggested by Littauer and Crouwel (1985, 94) for the chariot from the tomb of Amenhotep III. Tutankhamun's chariots with siding fill were interpreted by Littauer and Crouwel (1985, 73) as hunting chariots because they are far less sumptuously decorated and of a far lighter build:

"The large fenestrations shown on many chariots indicate that the siding fill in these cases cannot have been for protection, but was there to maintain tension on the two areas of the artificially bent-wood siding frame most in need of it, i.e. the centre front and the two rear verticals."

This means far less protection from the enemies' arrows, a situation that is, of course, not at all desirable or practical. Moreover, as noted by Littauer and Crouwel (1985, 70):

"the thong flooring [...] holds the bent-wood floor frame in shape and provides a resilient platform in an otherwise springless vehicle"

The thong flooring was covered with animal skin, sometimes in combination with a thick linen mat, or leather mat as in Yuya and Tjuiu's chariot (CG 51188). Thus, the choice of using leather for chariots rather than dressed thin wood, as seen in two chariots from Tutankhamun and the one from Thutmose IV seems obvious (Littauer & Crouwel 1985, 74):

"As compared with leather, a lighter and more resilient material, the relatively large wooden surfaces A1, A2 [the state chariots], the chariot of Yuya and Tuiu, and in particular that of Tuthmosis IV, would have tended to reduce the flexibility of the body and its ability to withstand strain. The material bespeaks a limited use, and this is confirmed by the elaborate decoration of these chariots" [text between [ ] inserted by present authors].

So, the material depended on the function of the chariot. Leather was a poor choice for a state chariot as it was not as costly a material as wood, and also, would have been harder to decorate with glass, semi-precious stones, and metal than wood.

But could there have been other reasons? Or alternatives? Clearly, the above-mentioned physical problems are the main reason for not using wood for certain types of chariots. Alternatives are rare. Textile, although possible as a chariot casing, would be impractical in the extreme, basketry and cartonnage chariots might be possible, albeit not for warfare, but there is no evidence for such constructions from Egypt.

Thus, leather chariots that weigh less than wooden ones and can therefore move faster and are more manoeuvrable would be more appropriate for hunting and even warfare than ones made of solid wood. Granted, wood would protect the person in the chariot better than leather, but thick leather might repel arrows and entangle spears a bit as well (see Cheshire, this volume). The floor of leather chariots (and even some wooden body chariots) was also made of leather strips that were plaited together, making the floor flexible and shock resistant; this required a strong sense of balance of the part of the passengers, but would also cushion them from jarring movements.

Rawhide and sinew were used in chariot production (the latter particularly in chariots with leather casing and in the other elements that belong to such chariots). Rawhide is very effective in chariot construction as it binds things together tightly and securely as damp rawhide shrinks upon drying: such properties were valuable wherever different portions of the frame and the body needed to be tightly joined, such as the overlap of the two parts of the floor frame, the joints in naves, spokes, and felloes, or as a tyre around the entire wheel. When rawhide was used in the construction of a chariot there was no fear of nails jolting loose at inopportune moments. This material was particularly ideal in Egypt where there was little chance of the rawhide becoming damp and thus coming loose.

Thin strips of sinew were often used to sew the leather together — this is particularly obvious in most chariots or fragments thereof that the EMCP has examined: sinew is used in construction as well as for securing decoration. The Tano chariot, however, is unusual in that flax was used to secure the elaborate decoration, and sinew was more commonly used for constructional joints or in areas that would experience more wear or stress.

Summary

The most important leather parts of some chariots are the casing, harness and floor. Rawhide and leather were also used as tyres and as lashing for securing various parts together. But the choice of using leather for the majority of the construction was dictated by the purpose of the chariot. This choice, as well as that of rawhide, has greatly helped to create the perfect chariot that has never been surpassed in lightness, swiftness, strength and riding qualities. Whether the finer details of using leather, rawhide, or sinew were inventions of the Egyptians or adaptations of established chariot technology that was introduced from the Near East is as yet unclear. Nonetheless, it is clear that the ancient Egyptians quickly adapted to using chariots in hunting and warfare, and started producing chariots for both ceremonial and practical purposes that were perfectly suited for their climate and terrain. This must have resulted in an increased importance of industries related to chariot production, such as wood and leather working.

References

Botti, B. (1951). Il Carro del sogno (per un grato ricordo personale). *Aegyptus* **31**. 192-198.

Carter, H. (1933). *The tomb of Tut.ankh.Amen. The annexe and treasury.* London: Cassel & Co.

Carter, H. & Newberry, P.E. (1904). *The Tomb of Thoutmosis IV.* Westminster: A. Constable and Co.

Daressy, M.G. (1902). *Catalogue général des antiquités Égyptiennes du musée du Caire. Fouilles de la Vallée des Rois.* Le Caire: L'Institut Francais d'Archéologie Orientale.

Davies, N. de Garis. (1963). *Scenes from some Theban tombs (Nos. 38, 66, 162, with excerpts from 81).* Oxford: Griffith Institute (Private Tombs at Thebes, Volume IV).

Forbes, R.J. (1957). *Studies in ancient technology.* Volume V. Leiden: Brill.

Guidotti, M.C. (ed.) (2002). *Il carro e le armi del museo Egizio di Firenze.* Firenze: Giunti Gruppo Editoriale.

Littauer, M.A. & Crouwel, J.H. (1985). *Chariots and related equipment from the tomb of Tutankhamuns.* Oxford: Griffith Institute.

Quibell, J.E. (1908). *Catalogue général Égyptiennes du musée du Caire. Nos 51001-51191. Tomb of Yuiaa and Thuiu.* Cairo: L'Institut Français d'Archéologie Orientale.

Veldmeijer, A.J. (2010). *Amarna's leatherwork. Part I. Preliminary analysis and catalogue.* Norg: DrukWare (republished in 2011 by Sidestone Press).

Veldmeijer, A.J. & Ikram, S. With a contribution by Skinner, L. (2011). Preliminary report of the Egyptian Museum Chariot Project (EMCP). *Bulletin of the American Research Center in Egypt.* 7-11.

Veldmeijer, A.J., Ikram, S. & Skinner, L. (2013). Charging chariots: progress report on the Tano chariot in the Egyptian Museum Cairo. In: Veldmeijer, A.J. & Ikram, S. (eds.). *Chasing chariots. Proceedings of the first international chariot conference (Cairo 2012).* Leiden: Sidestone Press. 257-271.

Why wineskins?

The exploration of a relationship between wine and skin containers

Barbara Wills & Amanda Watts

Introduction

The intention of this paper is to explore the history of wineskins; to investigate how animal skins might facilitate the production and transport of wine and how wine might alter the skin itself. Wineskins were in widespread use from ancient times; Homer's Odysseus, for example, makes Polyphemus drunk from the contents of skins. In 1st century AD Palestine a parable related by Jesus distinguishes used from unused wineskins:

> *"no one pours new wine into old wineskins. If he does, the new wine will burst the skins ... new wine must be poured into new wineskins" (Luke 5: 37-38).*

Wine was typically made by extraction of the grape juice followed by initial fermentation in a large container, straining off the lees and decanting into other vessels. Fermentation would evolve sufficient carbon dioxide to significantly stretch a skin if this were the chosen container, and the exothermic reaction (a possible temperature rise to above 30^{0}C) could begin to denature skin proteins. It might be possible to detect such changes in the archaeological record if (putative) ancient wineskins were found. Linings to leather containers such as the Spanish *bota*, made of tanned goatskin with an impermeable resin or pitch sealant, should also be detectable. The advantages and disadvantages of using skins to carry wine are considered: in terms of transportation, wineskins or leather bottles are less liable to breakage than ceramic or glass containers. They are also lightweight and conform well to animal and human transportation.

Ancient wine and viticulture

The first written record of winemaking comes from the Bible, when Noah, described as the first tiller of the soil, established a vineyard after the ark came to rest on Mount Ararat (The Holy Bible 1965, Genesis Ch. 9, V. 20-21). The discovery of what appears to be an early wine-making facility (in ancient Armenia, within 60 miles of Mount Ararat) dates back c. 6000 years (Barnard *et al.* 2010): grape seeds found were from *Vitis vinifera vinifera*, a grape still used to make wine today. Settlements (Hajji Firuz Tepe and Godin Tepe) in the northern Zagros Mountains have offered evidence of wine storage dating between 5000 and 5400 BC (Estreicher 2006, 11) uncovered during the archeological expedition of Mary M. Voigt from the University of Pennsylvania Museum. Thus an ancient wine culture, stimulated by the domestication of the Eurasian grape, was probably first established in the upland northern parts of the Middle East and subsequently extended to Egypt and the eastern Mediterranean (McGovern 1997, 3-7). However, more recent research suggests more than one origin of the domesticated grapevine (Arroyo-Garcia *et al.* 2006).

The most extensive early documentation about winemaking comes from Egypt (Murray *et al.* 2000). The Nile Delta has supported vineyards for five millennia and winemaking processes are represented on tomb walls dating to 2600 BC. Wine in ancient Egypt was a drink of great importance, consumed by the upper classes and the kings (Guash-Jané 2011). Even where beer was ubiquitous, special wines from the Nile Delta were required as funerary offerings, and much wine was drunk at major religious festivals. Wine, therefore, was often more than a drink to ancient societies, used widely as a ceremonial constituent in religious practice.

Pliny the Elder, author of Naturalis Historia (AD 23-79, a Roman of equestrian rank and considered the first encyclopaedist), writes of seven kinds of salted wines and eighteen varieties of sweet wines (Book XIV, Ch. 10 & 11). Wine itself, and additives to wine, were considered remedial: an example derived from garden plants is oak bark, which in wine is given to stem dysentery. Pliny the Elder writes (Book XIV, Ch. 24)

> *"the universal practice both there [Italy] as well as in the adjoining provinces [is] to season their new wines with resin".*

Certainly it was important to prevent 'wine disease': triterpenoids and diterpenoids from resins are effective in inhibiting the bacteria that produce acetic acid, so keeping wine from turning to vinegar. Herbs, spices, brine, oil and perfume were also added (Robinson 2006). In Roman (and pre-Roman) times wine was differentiated, valued and had a high commercial worth.

Ancient wineskins

Wine containers varied, and animal skins were only one of a range of materials that might be used. Ancient ceramics can testify to the presence of ancient wine, but where is the evidence for wineskins? In terms of artefacts, leather or skin rarely survives from ancient times unless preserved in exceptionally dry, or waterlogged anaerobic, conditions. During archaeological excavation, nondescript fragments of leather/skin may be overlooked, and wineskins can resemble other containers. Thus further investigation would be needed to confirm any association with wine. A water/wine skin from the second Jewish revolt was found in the Cave of

Letters (Schiffman *et al.* 1997). Exchanges regarding 140 skins on ostraca at Mons Claudianus seem to refer to waterskins alone (Daniel 1994) though fragmentary finds might possibly have carried wine. Markers that indicate wine include salts of tartaric acid, malvadins and, just possibly, tannins. Residues containing malvadin, the anthocyanin responsible for the red colour, however, point more clearly toward wine-making processes, deriving as they do from grape skins which are sometimes discarded early in the process.

Ancient literature refers to wineskins and some texts offer clues to processing. In the Biblical Septuagint, a young man addresses Job (32: 18-19):

> *"For I am full of words, the spirit within me constrains me. Behold, my heart is like wine that has no vent; like new wineskins, it is ready to burst".*

In the New Testament, the Apostles speaking in tongues say that they are full of must (Acts 2: 13). The impulse to deliver the spiritual message parallels the pressure created by fermentation, in both wineskin and belly (Immerwahr 1992, 130). In 1st century AD Palestine a parable related by Jesus distinguishes used from unused wineskins (Luke 5: 37-38):

> *"No one puts new wine into old wineskins, or else the new wine will burst the skins, and it will be spilled, and the skins will be destroyed. But new wine must be put into fresh wineskins, and both are preserved".*

Illustrations exist of wineskins, notably on Attic ceramics used in symposia. The skins have recognizable features; typically two or three bulbous legs (sometimes with carrying loops) and the neck shown in profile. Skins are variously depicted. A very large wineskin (British Museum 1805, 0703.458) is carted by oxen (figure 1). Decanting and mixing for feasting (such as the calyx-crater British Museum 1856, 1226.5) was a popular theme (figure 2). The skin was sometimes draped empty: Polyphemus lies drunk beside a draped skin (figure 3). It could be whirled around the head, or inflated as a plaything. Coinage also illustrates; the satyr Marsyas shoulders a wineskin on a silver denarius of 82 BC. Bacchanalian processions show wineskins; on a Roman sarcophagus from the Fitzwilliam Museum (GR.1.1835) a boy expertly pours the contents from a skin into a cup, and the same gallery shows a drinking scene with wineskin on a cup (GR.17.1937, c. 525-500 BC). Wineskins sometimes illustrate riotous play such as wine poured directly into the mouth of a satyr (figure 4). Satyrs and human both mounted blown-up skins (Museum of Fine Arts Boston No. 95.34). Silenus astride a wineskin becomes an oil lamp (British Museum Q3572) (Bailey 1996, 15). A game involving greased and inflated skins has been depicted; presumably those who kept their balance and stayed on longest won (Immerwahr 1992). There are a number of other wine-skin related activities listed in the Suda (Souda), a 10th century Byzantine encyclopaedic lexicon written in Greek and taking its sources from ancient writers, some via Medieval Christian compilers, as well as the Bible. These include the military ripping open wineskins filled with ash and soot to cloud the air (Suda On Line, "Portion, share"), various drinking contests, carrying wineskins in processions and comical dances involving hopping on greased wineskins (Suda On Line, "A wineskin in a frost").

Figure 1. Fragment of a 3rd c. AD Roman sarcophagus lid; relief of coarse marble (from Proconnesus or Thasos) showing the transport of wine being received at a gate. 1805,0703.458. © British Museum, London (see appendix).

Another, presumably secondary, use is as swimming aids; warriors cross a river clasping inflated skins seen in the Assyrian carved stone relief of Sennacherib's army (Nimrud palace panels British Museum 124541 and 124543, c. 865 - 860 BC).

Ancient winemaking

After harvest, grapes are crushed to release the juice. In ancient times these were sometimes contained in open-weave basketry bags (a column-krater in the Vatican Museum shows three sacks of grapes being trodden on a trestle) during pressing, and other methods also existed. The juice would either flow immediately into a closed container or remain steeping with the grape-skins and stalks. The tannins, anthocyanins, and potassium ions which later crystallize out as potassium hydrogen tartrate 'cream of tartar' then initiated a chemical reaction, producing the ester compounds which give smell and taste to the final wine. During fermentation sugar converts to alcohol by the action of yeast, and in warm climates this proceeds quickly and unavoidably. The exothermic reaction produces carbon dioxide, pressurising a container unless allowed to escape. The warmth generated (possibly exceeding 30^0 C) accelerates the yeast conversion process, allowing other bacteria to flourish and produce by-products such as lactobacter, which converts malic acid into lactic acid (giving a creamy mouth-feel to the final wine). Fermentation ceases either when all sugar has been converted, or the yeasts are killed. These die naturally in their alcoholic by-product at 14.5-15%, so it is probable that most

Figure 2. Calyx-krater of Dionysius feasting with maenads and a satyr. Apulian, 370-360 BC. 1856,1226.5. © British Museum, London (see appendix).

ancient wine was near this strength. In ancient times the variety and quantity of microbes accessing open vats could lead to less than predictable outcomes. Stems would impart their natural 2-methoxy-3-isopropylpyrazine vegetal aroma, and wild yeasts cause side reactions to produce some acetic acid and subtle vinegar flavours. The pH of the wine would be higher (nearer neutral) than in modern wines because of the potassium compounds derived from the grapeskins.

Advantages and disadvantages of wineskins

If skins were used as containers, at what point was the liquid added? Was grape juice poured in straight from the press, as young wine after initial fermentation or as the final product after maturation? It seems clear that skins were on occasion used at all of the above stages, though the skins differed according to function. What advantages might be conferred by fermenting or storing wine in skins or leather?

Figure 3. Calyx-krater showing Polyphemus drunk with empty wineskin draped alongside, Lucania circa 420-410BC. 1947,0714.18. © British Museum, London (see appendix).

A lightweight and flexible fermentation container

Goatskin is especially resilient and would offer a reliable, expandable container. Partial filling could have occurred as the winemaker knew the skin would subsequently inflate (Immerwahr 1992, 128) as the wine matured. Whether steps were taken to allow the excess carbon dioxide to escape is not known.

Regulation of temperature

A liquid in a skin that has not been lined with resin will slowly penetrate the skin to evaporate off on the surface, keeping the contents cool and stable.

Figure 4. Detail of psykter depicting excited satyr with wineskin. GR1868,0606.7. Caere, Italy, circa 500BC-470BC. © British Museum, London (see appendix).

Exclusion of oxygen and light

Wine in prolonged contact with oxygen will oxidise and spoil, and wine makers seek to exclude air. A skin will conform to its contents, and will deflate as wine is poured out, so can be 'collapsed' to exhale air and keep the wine stable. A modern parallel is the inner skin of a wine-box. Direct light can adversely react with phenolic compounds in the wine; a skin excludes light.

Separation from sediment

Potassium hydrogen tartrate is less soluble in alcoholic wine solutions than in non-alcoholic grape juice so as fermentation progresses it precipitates out to form small crystals the size of a sand grain. In ancient wines, this grainy sediment would have accumulated in the bottom of a vessel. When pouring, such crystals might cling to the inside of wineskins more readily, keeping the wine clearer than from a smooth rigid vessel.

Lessened agitation

Wine suffers from 'shock' if agitated: the subtle and volatile esters become detached from the longer molecule chains if too much energy is applied (from light, heat or motion). After a journey, wine needs to settle to allow these esters to reattach before they escape into the air. Wine crashes against a hard surface with more force than in a soft sided vessel, so skins may reduce agitation of the wine during transport.

Ease of transport

Traditional pitch-lined Spanish *botas* have been in use for centuries, an alternative to clay or pottery containers that can break during transportation (figure 5). During the late 1800s there was a boom of *bota* manufacturers in Spain which today barely survives (Winebotas, pers. comm. 2011). Both small containers for personal use and larger goatskins for storage are still used, albeit rarely (figure 6).

Large wineskins have been paralleled in modern times by the use of shipping containers. Bulk wine is placed in a conventional shipping container (a Flexitank) lined with an impermeable skin like that used in wine-boxes, the largest has a capacity of 24,000 litres. Wine is subsequently decanted into glass bottles for sale. According to WRAP (Waste and Resources Action Programme) (Hartley 2008, 3), shipping wine in bulk has both economic benefits and greater recycling capacity. It also improves product quality as bulk wine is less prone to temperature variations during transit due to thermal inertia.

Disadvantages

What are the disadvantages? Certainly wine may take on flavours from pitch-lined skins. The following passage, paraphrased from the observations of a Mr. Samuel Baker who visited Cyprus in 1879 (Rizopoulou-Egoumenidou 2009), is a vivid description:

> *"One mule carries two wineskins, one on each side. The vine-grower journeys over mountainous paths to Limassol to sell to the wine-merchant [...] the wine, contained in tarry goat-skins is, after a few hours exposure to the heat about the temperature of the hottest bath. He hurries his mules forward, in order to deliver the wine as quickly as possible before contamination by the skins [...] there are hundreds of proprietors who must be patient while their wine is imbibing the hateful flavour of the goatskins [...] at length the vine-grower offers a sample to the merchant; who, having spat it out, advises him to 'throw his wine into the sea, as it is undrinkable' having remained too long in the goat-skins exposed to the sun."*

Figure 5. Aragonese peasant with bota. Barbara Wills, after Kurt Hielscher.

Baker's views evidently were not shared by all. Wineskins continued in use at least until the mid-20th century. The tarry flavour was valued and considered wholesome by Cypriot natives, perhaps less so by those who had not grown up with it. Additionally, skin or leather containers may be pierced by sharp points, thus losing their contents.

Discussion

There is certainly evidence to show that skins were chosen to hold wine from ancient times, though they were of course not the only possible storage method. What features of archaeological leather would be sufficiently detectable to positively identify a putative wineskin? Markers that have successfully been used in ceramics to indicate wine contents include residual tartaric acid salts and malvadins. These may also aid the identification of wine in leather.

Figure 6. Modern bota made by Las Tres ZZZ, Pamplona, Spain. Photograph by Barbara Wills.

Wineskins that held an initial or secondary fermentation were changed by the process (we assume from the parable of Jesus) but in what way? The evolution of heat (up to 30^{0} C) and the increasing amount of alcohol would probably not have had an effect. Was the acidity of the wine significant? With a tanned skin, the pH parallels the acidity of the leather, but an oil-processed skin is closer to neutral so this might be affected by the acidity. Would it be possible to find wine-related tannins in the skin? Indeed, a red wine could be instrumental in vegetable-tanning a skin. It would therefore be important to understand the leather processing and tannage before applying analytical techniques so as to avoid any possible confusion with the contents. Perhaps the crucial thing here to look for would be the difference between the interior and exterior of the sample: more tannins on the inside and fewer on the outside might indicate wine contents.

Fermentation would stretch a skin significantly if the gasses built up, and the skin might eventually lose its ability to stretch further or to recover elasticity. Other liquids would also stretch a skin, but not to the same degree as fermenting must. Stretched skin is often clearly visible on examination, shown not only by the presence of stretch marks but from changes in skin diameter. Under the microscope a more parallel alignment of fibre bundles would be seen.

Conclusions

All this remains speculative until tested; the experimentation has not yet been done but it would be interesting to follow this up should resources become available to take this further. Based on knowledge of wine production and wineskins as explored through ethnographic, historical and practical sources, skins seem to have been favoured to carry wine on the basis of their properties as lightweight, flexible, watertight, repairable, re-usable containers.

References

Arroyo-García, R., Ruiz-García, L., Bolling, L., Ocete, R., López, M.A., Arnold, C., Ergul, A., Söylemezoglu, G., Uzun, H.I., Cabello, F., Ibáñez, J., Aradhya, M.K., Atanassov, A. Atanassov, I., Balint, S., Cenis, J.L., Costantini, L., Gorislavets, S., Grando, M.S., Klein, B.Y., McGovern, P.E., Merdinoglu, D., Pejic, I., Pelsy, F., Primikirios, N., Risovannaya, V., Roubelakis-Angelakis, K.A., Snoussi, H., Sotiri, P., Tamhankar, S., This, P., Troshin, L., Malpica, J.M., Lefort, F. & Martinez-Zapater, J.M., (2006). Multiple origins of cultivated grapevine *(Vitis vinifera L. ssp. sativa)* based on chloroplast DNA polymorphisms. *Molecular Ecology,* **15**. 3707-3714.

Bailey, D.M. (ed.) (1996). *A catalogue of lamps in the British Museum IV - Lamps of metal and stone, and lampstands.* London: British Museum Publications.

Barnard, H., Dooley, A.N., Areshian, G., Gaspary, B., Faulle, K.F. (2011). Chemical evidence for wine production around 4000 BCE in the Late Chalcolithic era. *Journal of Archaeological Science,* **38**, 5: 977-984.

Daniel, R. (1994). 'Neither do they put new wine into old skins'. *Zeitschrift für Papyrologie und Epigraphik,* **101**. 61-66.

Estreicher, S.K. (2006). *Wine from Neolithic times to the 21st century.* New York: Algora Publishing.

Guash-Jané, M.R. (2011). The meaning of wine in Egyptian tombs: the three amphorae in Tutankhamun's burial chamber. *Antiquity,* **85**, 329. 851-858.

Hartley, A. (2008). Bulk shipping of wine and its implications for product quality. Available at: http://www.wrap.org.uk/downloads/Bulk_shipping_wine_quality_May_08.5dfa7b6d.5386.pdf [accessed 14 June 2013].

Immerwahr, H. (1992). New wine in ancient wineskins: the evidence from Attic vases. *Hesperia, The Journal of the American School of Classical Studies at Athens,* **61**, 1. 121-132.

McGovern, P., Hartung, U., Badler, V.R., Glusker, D.L. & Exner. L.J., (1997). The beginnings of winemaking and viniculture in the ancient Near East and Egypt. *Expedition,* **39**, 1. 3-5.

Murray, M.-A., Boulton, N. & Heron. C., (2000). Viticulture and wine production. In: Nicholson, P. & Shaw. I., (eds.) *Ancient Egyptian materials and technology.* Cambridge: Cambridge University Pres. 577-578.

Pliny the Elder, The Natural History (English Translation). Bostock, J. & Riley. H.T., (eds.). Book XIV. The natural history of the fruit trees. Available online through The Persus Catalogue: http://www.perseus.tufts.edu/hopper/text?doc=Perseus:text:1999.02.0137 [Accessed 19/02/2014].

Rizopoulou-Egoumenidou, E. (2009). *Tanning in Cyprus from the 16th to the 20th century. From traditional tanneries to modern industries.* Nicosia: Cyprus Research Centre. 129-130.

Robinson, J. (2006). *The Oxford companion to wine.* 3rd edition. Oxford: Oxford University Press.

Schiffman, L.H. , Tov, E. & Van der Kam, J.C., (eds). (1997). *The Dead Sea scrolls: fifty years after their discovery. Proceedings of the Jerusalem Congress, July 20-25.* Jerusalem, Israel Exploration Society/The Shrine of the Book, Israel Museum.

Suda On Line, *Portion, share.* Trans. Benedict, J. 14 March 2002. Available at http://www.stoa.org/sol-entries/alpha/3445. Published by the Stoa Consortium [Accessed 17 January 2014].

Suda On Line, *A wineskin in a frost.* Trans. Benedict, J. 30 March 2002. Available aium [Accessed 17 January 2014].

The Holy Bible (1965). Revised Standard Version Catholic Edition. London: Catholic Truth Society. J.C.

Appendix

Figure 1:

http://www.britishmuseum.org/research/collection_online/collection_object_details.aspx?objectId=399723&partId=1&searchText=1805,0703.458&page=1

Figure 2:

http://www.britishmuseum.org/research/collection_online/collection_object_details.aspx?objectId=463173&partId=1&searchText=1856,1226.5&page=1

Figure 3:

http://www.britishmuseum.org/research/collection_online/collection_object_details.aspx?objectId=463183&partId=1&searchText=1947,0714.18&page=1

Figure 4:

http://www.britishmuseum.org/research/collection_online/collection_object_details.aspx?objectId=461894&partId=1&searchText=1868,0606.7&page=1